Project Earth Science:
Geology

Excerpt from "Sensations and Reflections Caused by the Earthquake in January Last" by Flaccus from THE KNICKERBOCKER OR NEW-YORK MONTHLY MAGAZINE, volume 18, pp. 27–28, 1841.

"Orbiter 5 Shows How the Earth Looks from the Moon" by May Swenson. Reprinted with the permission of Simon & Schuster Books for Young Readers, an imprint of Simon & Schuster Children's Publishing Division from THE COMPLETE POEMS TO SOLVE by May Swenson. Copyright © 1993 The Literary Estate of May Swenson.

"Goodnight" from IN THE STONEWORK by John Ciardi. Copyright © 1961 by the Ciardi Family Publishing Trust. Reprinted by permission of the Ciardi family.

Excerpt from "Rock" from THE YEAR ONE by Kathleen Raine. Reprinted by permission of the author.

"Earthquake" by Kokan Shiren from FROM THE COUNTRY OF EIGHT ISLANDS by Hiroaki Sato and Burton Watson. Copyright © 1981 by Hiroaki Sato and Burton Watson. Used by permission of Doubleday, a division of Bantam Doubleday Dell Publishing Group, Inc.

Excerpt from "How the ground escapes me" from CONTEMPORARY SPANISH POETRY, p. 395, translated by Eleanor L. Turnbull. Copyright © 1945 by The Johns Hopkins Press. Reprinted by permission of The Johns Hopkins University Press.

Excerpt from "What Happened Here Before" from TURTLE ISLAND by Gary Snyder. Copyright © 1974 by Gary Snyder. Reprinted by permission of New Directions Publishing Corp.

Excerpt from THIS SCULPTURED EARTH by John Shimer. Copyright © 1959 by Columbia University Press. Reprinted with permission of the publisher.

"Born Was the Island" from FORNANDER COLLECTION OF HAWAIIAN ANTIQUITIES AND FOLK-LORE by T.G. Thrum. Honolulu: Bishop Museum Mem., 4–6, 1916.

"Mount Fuji, Opus 5" by Kusano Shimpei from THE PENGUIN BOOK OF JAPANESE VERSE, translated by Geoffrey Bownas and Anthony Thwaite (Penguin Books, 1964, p. 216). Copyright © 1964 Geoffrey Bownas and Anthony Thwaite. Reprinted by permission of Penguin Books Ltd.

"The Santa Barbara Earthquake" sung by Vester Whitworth, from Library of Congress record AFS 4098 B1. Recorded by Charles Todd and Robert Sonkin, Arvin, California, 1940.

Excerpted from THE THOUGHTS OF THOREAU By Henry David Thoreau. New York: Dodd, Mead & Company, 1962.

Excerpt from "The River Duddon, Sonnet XV" from COMPLETE POETICAL WORKS by William Wordsworth, 1888.

Excerpt from "A Poetical Geognosy" from A GEOLOGICAL PRIMER IN VERSE: WITH, A POETICAL GEOGNOSY; OR FEASTING AND FIGHTING, AND SUNDRY OTHER RIGHT PLEASANT POEMS WITH NOTES by John Scafe. London: Longman, Hurst, Rees, Orme, and Brown, 61 pp., 1820.

Library of Congress Card Catalog Number 95-067463

Stock Number PB111X   ISBN 0-87355-131-1

♻ Printed in the United States of America on recycled paper.

The National Science Teachers Association is an organization of science education professionals and has as its purpose the stimulation, improvement, and coordination of science teaching and learning.

Project Earth Science: Geology

by

Brent A. Ford

A Project of Horizon Research, Inc.
Material for middle school teachers in Earth science
This project was funded by BP America, Inc. and
published by the National Science Teachers Association

Project Earth Science: Geology

Table of Contents

Activities

NATIONAL SCIENCE TEACHERS ASSOCIATION

H. Weller.

Readings

Appendices

About the cover images

Front Cover

• Mount St. Helens dome thermal map compiled by NASA's Jet Propulsion Laboratory. This airborne remote sensing survey depicts the lava dome as it appeared on September 1, 1988, and was created by registering the image data to a 1:4000 scale digital elevation model. Hot spots appear bright blue, and are the result of fumarolic activity. Downloaded off the Internet courtesy NASA/JPL.

• Walcott Tuff exposed in a cliff along the Snake River near Massacre Rocks State Park, Idaho. Photo courtesy Mike Horn, Centennial High School, Meridian, Idaho.

Back Cover

• GIF image of crustal ages of Earth's seafloor using data compiled by the Scripps Institution of Oceanography. Downloaded off the Internet courtesy U.S. National Geophysical Data Center/ World Data Center A for MGG.

H. Weller.

Acknowledgments

Numerous people have contributed to the development of _Project Earth Science: Geology_. The volume began as a collection of activities and readings from Project Earth Science, a teacher enhancement program funded by the National Science Foundation. Project Earth Science was designed to provide in-service education for middle school Earth science teachers in North Carolina. Nine two-person leadership teams received extensive training for conducting in-service workshops on selected topics in astronomy, meteorology, physical oceanography, and geology. They, in turn, conducted in-service education programs for teachers throughout North Carolina. Principal investigators for this project were Iris R. Weiss, president of Horizon Research, Inc.; Diana L. Montgomery, research associate at Horizon Research, Inc.; Paul B. Hounshell, professor of education, UNC-Chapel Hill; and Paul D. Fullagar, professor of geology, UNC-Chapel Hill.

The activities and readings in this book have undergone many revisions as a result of the comments and suggestions provided primarily by the Project Earth Science leaders, and also by workshop participants, project consultants, and project staff. The Project Earth Science leaders made this book possible through their creativity and unceasing attention to the needs of students and classroom teachers. The leaders were Kevin Barnard, Winston-Salem/Forsyth County Schools; Kathy Bobay, Charlotte-Mecklenburg Schools; Pam Bookout, Guilford County Schools; Betty Dean, Guilford County Schools; Lynanne (Missy) Gabriel, Charlotte-Mecklenburg Schools; Flo Gullickson, Guilford County Schools; Michele Heath, Chapel Hill/Carrboro Schools; Cameron Holbrook, Winston-Salem/Forsyth County Schools; Linda Hollingsworth, Randolph County Schools; Geoff Holt, Wake County Schools; Kim Kelly, Chapel Hill/Carrboro Schools; Laura Kolb, Wake County Schools; Karen Kozel, Durham County Schools; Kim Taylor, Durham County Schools; Dana White, Wake County Schools; Tammy Williams, Guilford County Schools; and Lowell Zeigler, Wake County Schools.

Special thanks to the following: Geoff Holt for his contributions to "Rock Around the Clock"; Dr. Dirk Frankenberg for his contributions to "Edible Tectonics"; Robert C. Lagemann for writing and illustrating three of the readings; and Linda Ford, vice president of Novostar Designs, Inc., for significant contributions to the readings and many of the activities.

Thanks to Kim Taylor, who contributed significantly in the preparation of the bibliography, and to Kevin Barnard, Kathy Bobay, Missy Gabriel, Linda Hollingsworth, and Tammy Williams, who assisted in annotating the citations. The manuscript was reviewed for accuracy by Paul Fullagar, professor of geology at UNC-Chapel Hill. Dr. Fullagar contributed valuable suggestions to all the activities and readings. The manuscript was also reviewed by Shiela Marshall, NSTA; Jeff Callister, Newburgh Free Academy; Sharon Stroud, Widefield High School; Cheryl Zielaskowski, Washington Middle School; Janet Woerner, California State University-San Bernardino; Katherine Becker, Creighton University; Len Sharp, Liverpool High School; Keith Sverdrup, University of Wisconsin-Milwaulkee; Martin Stout, California State University–Los Angeles; Frank Ireton, American Geophysical Union; David Simpson, Incorporated Research Institute for Seismology (IRIS); and Lucy Jones, U.S. Geological Survey Seismological Laboratory.

Shirley Brown, a teacher with the Columbus City Schools in Ohio and a participant in the Program for Leadership in Earth Systems Education (PLESE) at Ohio State University, contributed many of the ideas for the "Suggestions for Interdisciplinary Study" sections.

Project Earth Science: Geology is published by NSTA—Gerry Wheeler (Executive Director); Phyllis Marcuccio (Executive Director for Publications). NSTA Special Publications produced *Project Earth Science: Geology*—Shirley Watt Ireton (Managing Editor); Chris Findlay and Joe Cain (Project Editors); Gregg Sekscienski and Doug Messier (Associate Editors); Michelle Eugeni (Assistant Editor); and Jennifer Hester (Editorial Assistant). Max-Karl Winkler drew the illustrations from sketches created by the author. Marty Ittner of Auras Design, Inc., designed the cover. The book was printed by Automated Graphic Systems.

Special thanks go to BP America for funding the publication of the Project Earth Science series, and to the National Science Foundation (grant number TPE-8954625) for funding the Project Earth Science teacher enhancement project. This publication does not necessarily reflect the views of BP America or the National Science Foundation.

Introduction

Project Earth Science: Geology is the fourth of the four-volume Project Earth Science series. The other three volumes in the series are *Astronomy*, *Meteorology*, and *Physical Oceanography*. Each volume contains a collection of hands-on activities and a series of readings developed for middle-level students.

Overview of Project Earth Science

Project Earth Science was a teacher enhancement program funded by the National Science Foundation. Originally conceived as a program in leadership development, this project was created to prepare middle-school science teachers to lead workshops on topics in Earth science. Workshops helped teachers learn to convey key Earth science concepts and content through the use of hands-on activities in the classroom. With the help of content experts, concept outlines have been developed for specific topic areas, and activities have been designed to illustrate those concepts. Several of the activities in this volume draw directly from existing sources; the remainder have been developed by Project Earth Science leaders and staff. Over the course of this two-year program, activities were field tested in teacher workshops and classrooms. Activities have significantly improved as a result of detailed participant evaluation. The effectiveness and appropriateness of activities have been reviewed by content experts, and organized into a standardized format. During the publication process these activities have again undergone extensive review.

About *Project Earth Science: Geology*

Project Earth Science: Geology is built upon the unifying theory of plate tectonics and explores how this concept can be used to explain the occurrences of volcanoes, earthquakes, and other geologic phenomena. It also provides a link between plate tectonics, rock and mineral types, and the rock cycle. Integrated into this foundation are a variety of points regarding the process of scientific investigation and modeling. The intent is to increase student awareness of how scientific knowledge is created.

This book is divided into three sections: activities, readings, and appendices. The activities are constructed around several basic concept divisions. First, understanding that Earth's surface is composed of plates that can move independently of one another is central for appreciating how plate tectonics occurs.

Second, Earth's tectonic plates move because they "ride" on a rock layer called the asthenosphere. The rocks comprising the asthenosphere have properties different from those comprising the plates. Several activities introduce students to these differences and encourage them to consider the driving forces for plate motion. Third, directly observing tectonic motion and mechanisms is difficult. Usually, geologists study these subjects indirectly by investigating a variety of geologic features and events on and near Earth's surface. Finally, the rocks and minerals around us today are products of complex geological processes. Evidence gleaned from their close study provides insights into the geological processes to which they have been subjected.

An understanding of the concept of density is required for several of the activities contained in this volume. The activities are written with the assumption that students have this understanding. If students have not yet learned this concept, there are several activities in *Project Earth Science: Physical Oceanography* and in *Project Earth Science: Meteorology* that can be used prior to conducting the activities herein.

A series of overview readings supports these activities. By elaborating concepts presented in the activities the readings are intended to enhance teacher preparation and serve as additional resources for students. The readings also introduce supplemental topics so teachers can link contemporary science to broader subjects and other disciplines.

For those interested in using this volume in teacher workshops, a Master Materials list is provided as an appendix. Also included is an Annotated Bibliography that serves as a supplemental materials guide. Entries are subdivided into various categories: *Activities and Curriculum Projects*; *Books and Booklets*; *Audiovisual Materials*; *Instructional Aids*; *Information and References*; *State Resources*; and *Internet Resources*. Although far from exhaustive, this compilation offers a wide range of instructional materials for all grade levels.

Creating Scientific Knowledge

Investigating plate tectonics offers a superb opportunity to encourage student thinking on two subjects: how scientific knowledge is created and how scientific knowledge evolves. The theory of plate tectonics gained wide acceptance only in the 1960s. Its implications produced a revolution within geology, forcing a complete reshaping of basic geological theories. The effects of this revolution are still being explored. *Project Earth Science: Geology* presents a variety of opportunities for teachers to discuss the creation and evolution of scientific knowledge. For example, students might consider

- how models channel—yet sometimes restrict—our conceptions of nature
- how scientific knowledge changes over time
- how our choice of measurement scale affects our perceptions of nature and of change

Models and analogies are extremely effective tools in scientific investigation, especially when the subject under study proves to be too large, too small, or too inacessible for direct study. Although models are used widely by Earth scientists, students must be reminded that models are not perfect representations of the object or phenomenon under study. It is essential that students learn to evaluate models for strengths and weaknesses, such as aspects in which models accurately represent phenomena and aspects in which they do not. Misconceptions about geological processes can be introduced when models are used beyond their range of application. Teachers can challenge misconceptions when they arise by discussing a model's advantages and limitations.

As students learn science, it is easy for them to lose sight of the fact that scientific knowledge evolves. As scientists gather more data, test hypotheses, and develop increasingly sophisticated means of investigation their understanding of natural phenomena changes. The theory of plate tectonics, for example, has evolved significantly over the last century.

"Continental drift" was first proposed in the early 1900s to explain why the outlines of continents seemed to fit so well together—the western coast of Africa and the eastern coast of South America, for example—and to explain why fossils found in rocks on the coasts of different continents were so similar. Having no mechanism to explain how such "drifting" could occur, most geologists rejected these early proposals in favor of alternative explanations. In the 1950s, geologists studying patterns in the frequency and location of volcanoes and earthquakes suggested that continents were imbedded in enormous "plates" of Earth's crust. As these plates shifted, they proposed, land areas near the shifts were jolted (earthquakes), and openings between the plates allowed magma to rise (producing volcanoes). Geophysicists remained skeptical of this model, however, because they still were unsure about what might drive such a process. At the same time, competing explanations accounted for the available data.

In the 1950s and 1960s, much research was performed

concerning the seafloor. Geologists found patterns in the relative ages and the magnetic orientations of rock formations on the seafloor. These patterns, combined with newer seismic studies of Earth's interior, provided compelling evidence that Earth's surface is composed of moving plates and that the continents ride atop them. Theories of how this could occur became more sophisticated and, since the late 1960s, plate tectonics has been accepted by virtually all Earth scientists as an accurate account of how Earth's surface changes over time.

With growing information and expanded understanding, scientific knowledge changes; what seemed absurd to many at the start of the century is accepted at century's end. Teachers should emphasize this changing nature of science—it is what makes scientific inquiry special as a form of knowledge—and encourage students to investigate in more detail how scientific knowledge evolves.

Observing and the Problem of Scale

Central to understanding the evolving character of science is appreciating the limits of our perceptions of change. We observe the world as it *is*, and our thoughts about how it *was* and how it *could be* tend to be quite restricted. That our world is changing constantly can be a difficult concept for students to accept. In several respects, this is a function of the rate at which change sometimes occurs compared to the length of time available to humans for direct observation.

To illustrate this point, ask students to consider the life of an insect that spends its entire existence—from June to August of a single year—in an oak tree. As outside observers, humans can observe seasonal and annual changes in the tree's biology. Because of the relatively short duration of its life, the insect cannot observe these changes.

Likewise, due to the relatively short span of our lifetimes compared to geologic time, people often have difficulty appreciating the changes taking place on a million-year scale. Continents move at a rate of less than ten centimeters per year; average global temperatures may change only a few degrees over thousands of years; mountain ranges can take millions of years to rise. Changes such as these are almost imperceptible during a person's life span. It is important for students to understand that while observing these changes may be difficult, Earth's geologic features are continually changing. Comparing events and changes on different scales often is a difficult skill for students to acquire.

Moreover, diagrams and models will exaggerate or compress relative sizes to make a certain point more obvious. Such descriptions, however, usually change one scale in the representation while ignoring others. A common example involves displays of our solar system that accurately depict the relative *distances* between planets but misrepresent the planets' relative *sizes*. It is important for teachers to discuss the concept of scale and encourage students to raise questions about the various measurement scales used in these activities.

Getting Ready for Classroom Instruction

The activities in this volume are designed to be hands-on. In collecting and developing them, effort was made to use materials that are either readily available in the classroom or inexpensive to purchase.

Each activity has two sections: a Student section and a Teachers Guide. Each student section begins with *Background* information to briefly explain, in non-technical terms, the concepts involved in the activity. Following this introduction is a step-by-step *Procedure* outline and a set of *Questions/ Conclusions* to facilitate student understanding, encourage constructive thinking, and advance the drawing of scientific conclusions. Each activity begins with a poem or quote whose imagery has a theme related to the geological phenomenon under consideration. Teachers may choose to present these poems to students when activities are introduced. This practice will reinforce the point that the arts and sciences are interwoven.

The Teachers Guide contains a more thorough version of the background information given to students, plus a summary of *Important Points for Students to Understand* in the activity. *Time Management* estimates the duration of each activity, and *Preparation* describes the set-up and lists sources of materials for some. To challenge students to extend their study of each topic, *Suggestions for Further Study* are provided. For relating the science in each activity to other disciplines, such as language arts, history, and social sciences, *Suggestions for Interdisciplinary Reading and Study* are provided, as are subject-specific poems at the beginning of each activity. The final portion of each Teachers Guide provides *Answers to Questions for Students*.

Although the scientific method has often been presented as a "cookbook" recipe—state the problem, gather information, form a hypothesis, perform experiments, record and analyze data, and state conclusions—students should be made aware that the scientific method provides an approach to understanding the world around us, an approach that is rarely so straightforward. For instance, many factors can influence experimental outcomes, measurement precision, and the reliability of results. Such variables must be taken into consideration throughout the course of an investigation.

As students work through the activities in this volume, they should be made aware that experimental outcomes can vary and that repetition of trials is important for developing an accurate picture of concepts under study. By repeating experimental procedures, students can learn to distinguish between significant and insignificant variations in outcomes. Regardless of how carefully an experiment is conducted, error can never be entirely eliminated. As a matter of course, students should be encouraged to look for ways to eliminate sources of error. However, they also must be made aware of the inherent variation possible in all experimentation.

Finally, control of variables is important in maintaining the integrity of an experiment. Misleading results and incorrect conclusions often can be traced to experimentation where important variables are not rigorously controlled. Teachers should encourage students to identify experimental controls and consider the relationships between the variables under study and the factors held under control.

Key Concepts

The Activities in this book are organized within the context of the unifying geological theory of plate tectonics, which encompasses the following four key concepts. First is that Earth's surface, the lithosphere, is comprised of individual tectonic plates. Second, the plates move or "ride" atop the asthenosphere, and the nature of their motion (i.e., the direction) is determined by geological processes within Earth. Third, many geological events and features occurring on Earth's surface, such as volcanoes, earthquakes, and land formations, are directly related to plate tectonic activity. Finally, the rocks and minerals that comprise such features are produced over time by complex, cyclical geological processes.

Project Earth Science: Geology and the National Science Education Standards

Effective science teaching within the middle-level age cluster integrates the two broadest groupings of scientific activity identified by the *National Science Education Standards:* (1) developing skills and abilities necessary to perform scientific inquiry, and (2) developing an understanding of the implications and applications of scientific inquiry. Within the context of these two broad groupings, the Standards identify specific categories of classroom activity that will encourage and enable students to integrate skills and abilities with understanding.

To facilitate this integration, an organizational matrix for *Project Earth Science: Geology* appears on pages 16 and 17. The categories listed along the X-axis of the matrix, listed below, correspond to the categories of performing and understanding scientific activity identified as appropriate by the Standards.

Subject Matter and Content. Specifies the geological topic covered by an Activity.

Scientific Inquiry. Identifies the "processes of science" (i.e., scientific reasoning, critical thinking, conducting investigations, formulating hypotheses) employed by an Activity.

Unifying Concepts and Processes. Links an Activity's specific geological topic with "the big picture" of scientific ideas (i.e., how data collection techniques inform interpretation and analysis).

Technology. Establishes a connection between the natural and designed worlds.

Personal/Social Perspectives. Locates the specific geological topic covered by an Activity within an accessible framework.

Historical Context. Portrays scientific endeavor as an ongoing human enterprise by linking an Activity's topic with the evolution of its underlying principle.

By integrating the presentation of specific science subject matter with the encouragement of students to organize and locate that subject matter within an accessible framework, *Project Earth Science: Geology* hopes to address the *National Science Education Standards'* call for making science—in this case geology—something students do, not something that is done to students. The Organizational Matrix on the following pages provides a tool to assist teachers in realizing this goal.

Activity #	Subject Matter and Content	Scientific Inquiry	Unifying Concepts and Processes
Activity 1	global earthquake distribution	working with data	patterns in nature
Activity 2	Earth's geologic layers	evaluating strengths and weaknesses	modeling
Activity 3	plate tectonics	visualizing complex natural phenomena	unifying theories
Activity 4	continental movement	relating evidence to theory	forming hypotheses
Activity 5	behavior of the asthenosphere	comparing and contrasting	change and constancy
Activity 6	what causes continental movement	experimentation	modeling
Activity 7	mapping the ocean floor	using tools and techniques to gather data	"real-world" limitations on scientific undertakings
Activity 8	undersea geologic features	conceptualizing complex processes	modeling
Activity 9	why volcanoes form	investigation	understanding relationships
Activity 10	volcanic emissions	observation	modeling
Activity 11	why volcanic island chains form	thinking critically about evidence and explanation	modeling
Activity 12	what causes earthquakes	experimentation	measuring natural phenomena
Activity 13	the rock cycle	understanding scientific concepts	understanding change over time
Activity 14	rock formations	sampling	describing
Activity 15	rock identification	observing specimen characteristics	extrapolating from representative specimens

Technology	Personal/Social Perspectives	Historical Context	Activity Name
	earthquakes affect many people around the world	scientific data can represent decades of activity	**GeoPatterns**
		historical perspectives of Earth	**All Cracked Up**
		scientists don't always agree	**Edible Tectonics**
animating long-term natural processes		scientific theories evolve	**A Voyage Through Time**
			Solid or Liquid?
convection's use in human invention			**Convection**
using technology in scientific experimentation	prioritizing resources	new techniques impact old theories	**A Drop in the Bucket**
technological advances alter human perceptions of nature		science is a human endeavor	**Seafloor Spreading**
	natural hazards		**Volcanoes and Plates**
human uses of volcanic products	cultural interpretations of volcanic activity		**Volcanoes and Magma**
	natural resources	technological advances provide new data	**Volcanoes and Hot Spots**
constructing quakeproof buildings	cost and risk benefit analysis	comparing historical earthquakes	**Shake It Up**
human uses of "Earth materials"	types of natural resources	understanding geologic time	**Rock Around the Clock**
technological advances affect scientific theories	local and regional land formations		**Study Your Sandwich, & Eat It Too!**
	the rocks around us	creating narratives is important in science	**Rocks Tell a Story**

from *Sensations and Reflections Caused by the Earthquake in January Last*

...what double terror then
When sober Earth mimics the reeling sea!
And plains, upheaving into billows, yield
Unsolid to the foot of man and beast;
When our sure dwelling, like a foundering bark,
Pitches and rolls, the plaything of those strange
Unnatural waves, while hideous underneath
Yawn greedier caves than deepest ocean hides,
Glutted with fragments of the shipwrecked earth,
Clashing and plunging down! O!

Flaccus

GeoPatterns

Background

One of the main things scientists do is look for **patterns**, especially patterns in nature. Scientists look for patterns so they can answer questions about the way the universe works. Why do animals have certain kinds of markings? Why do particular plants grow in certain ways? Why do some planets have rings around them and others do not? Why do earthquakes occur more often in some places than in others? Explaining why patterns like these exist is one of the main goals of scientific research.

Take earthquakes, for example. In the United States, earthquakes occur often in California and Alaska, but not very often in Iowa and Nebraska. Why do earthquakes occur more often in California than in Nebraska? Do earthquakes occur randomly, or are there patterns to their distribution, to where they most often occur around the world? What causes earthquakes? What determines where an earthquake will occur? Geologists have looked for answers to these kinds of questions for decades. If they can predict when and where an earthquake will occur, they can warn people and prevent loss of life and damage to property.

In this Activity you will investigate some earthquakes that have occurred in different parts of the world. You will look for patterns—or the absence of patterns—in the frequency and distribution of earthquakes around the world. You will then try to explain your observations.

Objective

To study earthquake frequency and distribution around the world, look for patterns, and offer explanations for what you observe.

Materials

Each student or group of students will need
◊ all five panels of the strip map
◊ scissors
◊ glue (or clear tape)
◊ map of the ocean floor
◊ colored pencils (optional)
◊ atlas

Vocabulary

Pattern: A design; a consistent relationship.

Procedure

1. Your teacher will give you five strip map panels. Cut the panels out along their outside edges.

2. Lay the panels face up on your table, aligning them as shown in Diagram 1. Glue (or tape) the panels together along the tab lines, but at this point *do not connect* panel 5 to panel 1.

3. Your teacher may suggest that you color certain regions of the map. It is easiest to color your panels at this point, before you transform them into a globe.

4. Use an atlas to update the epicenter data by adding marks for each earthquake listed in the Data Table (page 21).

5. Once the coloring is complete, fold each of the panels along the straight lines printed on the map and along the tab lines. (This will make forming the globe shape easier.)

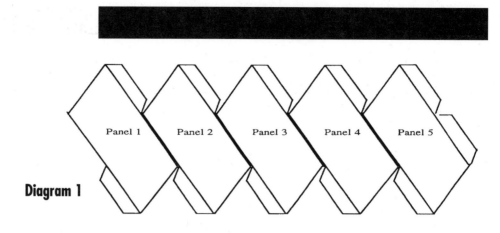

Diagram 1

6. Join panels 5 and 1 by gluing or taping together their tabs. Finish your globe by folding panels inward along the creases and gluing or taping connecting panels together.

7. Once you are finished, examine your globe closely. Locate the various continents and ocean basins. Your globe also shows thousands of dots. Each dot marks the location on the surface directly above where an earthquake originated (its epicenter). Look closely at the distribution of these dots over Earth's surface.

8. Use your globe and other resources in your classroom to answer the questions below. When you have completed this Activity, hang your globe nearby. You can use it later for other Activities.

Questions/Conclusions

1. What do you observe about the location of earthquake epicenters on your globe? Are earthquakes distributed randomly or in a pattern around the world?

2. Does the earthquake data from the Data Table add to or detract from any patterns you observed?

3. Why do you think earthquakes occur where they do?

4. Active volcanoes appear in locations often, but not always, near earthquake activity. Is this a coincidence or could earthquake and volcanic activity be related?

5. Examine a map of Earth's seafloor. What do you observe?

6. Do certain features appear in a pattern, or does randomness seem to be the rule on the seafloor? Provide some possible explanations for your observations.

7. Compare the distribution of earthquakes on your globe and the distribution of features on the seafloor. Do you think there is a connection between these different phenomena?

DATA TABLE: RECENT EARTHQUAKE EPICENTERS, LOCATIONS, AND MAGNITUDES

EPICENTER DATE	LOCATION	MAGNITUDE (Richter)
May 14, 1995	northern Greece	6.6
Jan. 17, 1995	Kobe, Japan	6.9
May 28, 1995	Russian coast	7.5
July 4, 1994	southern Mexico	6.0
June 8, 1994	Bolivia	8.2
Jan. 17, 1994	Los Angeles, California	6.8
Sept. 30, 1993	southern India	6.4
Aug. 8, 1993	Agana, Guam	8.0
July 12, 1993	ocean off Hokkaido, Japan	7.8
March 26, 1993	southern Greece	5.1
March 12, 1993	Fiji	6.5
March 6, 1993	Solomon Islands	6.5
March 6, 1993	Fiji	6.7
March 6, 1993	Santa Cruz Islands	7.1
Jan. 15, 1993	Kushiro, Japan	7.0
Dec. 20, 1992	Banda Sea, Indonesia	7.0
Dec. 12, 1992	Flores, Indonesia	7.5
Nov. 8, 1992	Fiji	6.5
Oct. 23, 1992	Papua, New Guinea	6.7
Oct. 17, 1992	Murinndo, Colombia	7.2
Oct. 12, 1992	Cairo, Egypt	5.9
Sept. 11, 1992	southeastern Zaire	6.8
Sept. 2, 1992	ocean off Nicaragua	7.2
Aug. 19, 1992	Kyrgyzstan	7.5
Aug. 7, 1992	Gulf of Alaska	6.5
June 28, 1992	southern California	7.5
May 25, 1992	Cabo Cruz, Cuba	7.0
May 15, 1992	Kyrgyzstan	6.2
April 25, 1992	northern California	7.0
April 13, 1992	southeastern Netherlands	5.0
March 13, 1992	Turkey	6.2
Feb. 27, 1992	eastern New Guinea	6.7

From the *1995 Information Please World Almanac*

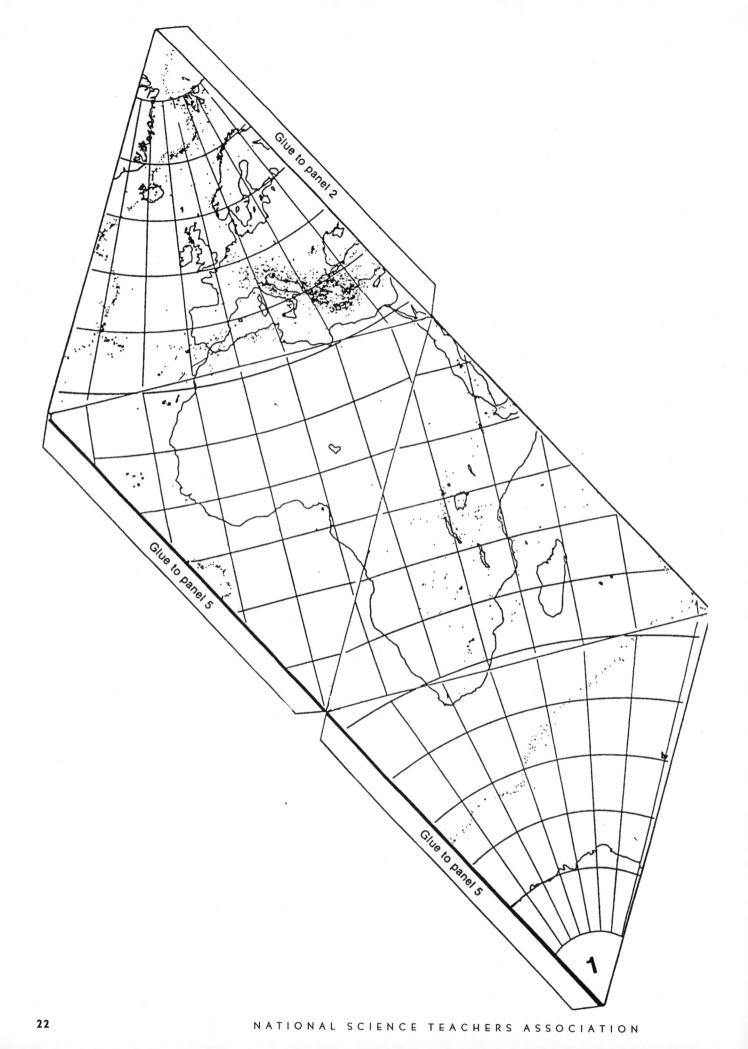

Glue to panel 2

Glue to panel 5

Glue to panel 5

1

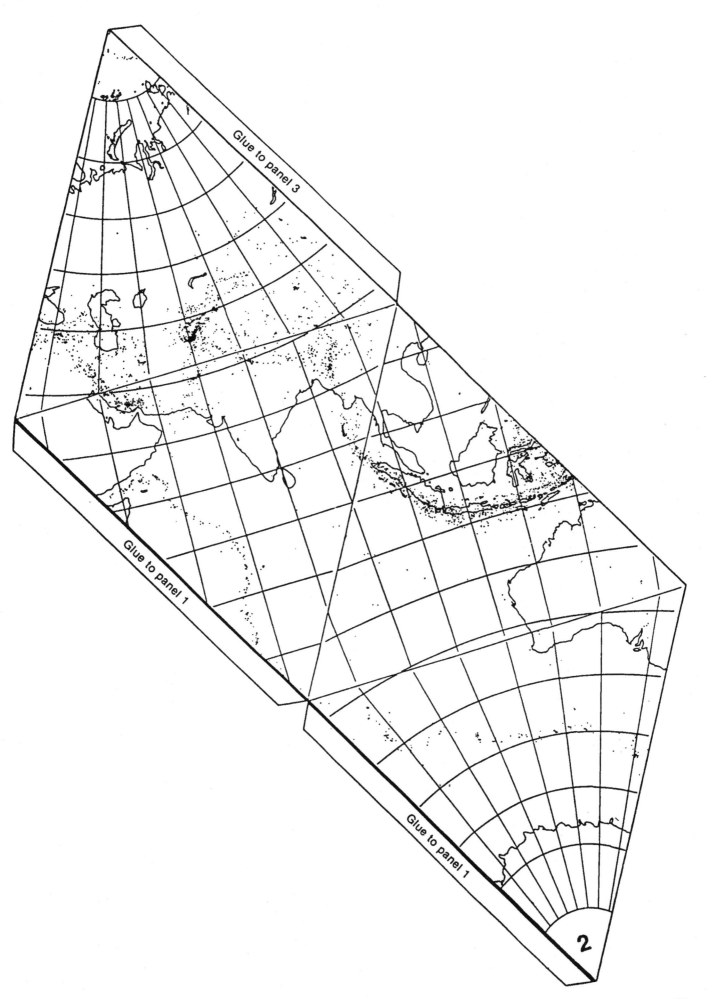

Glue to panel 3

Glue to panel 1

Glue to panel 1

2

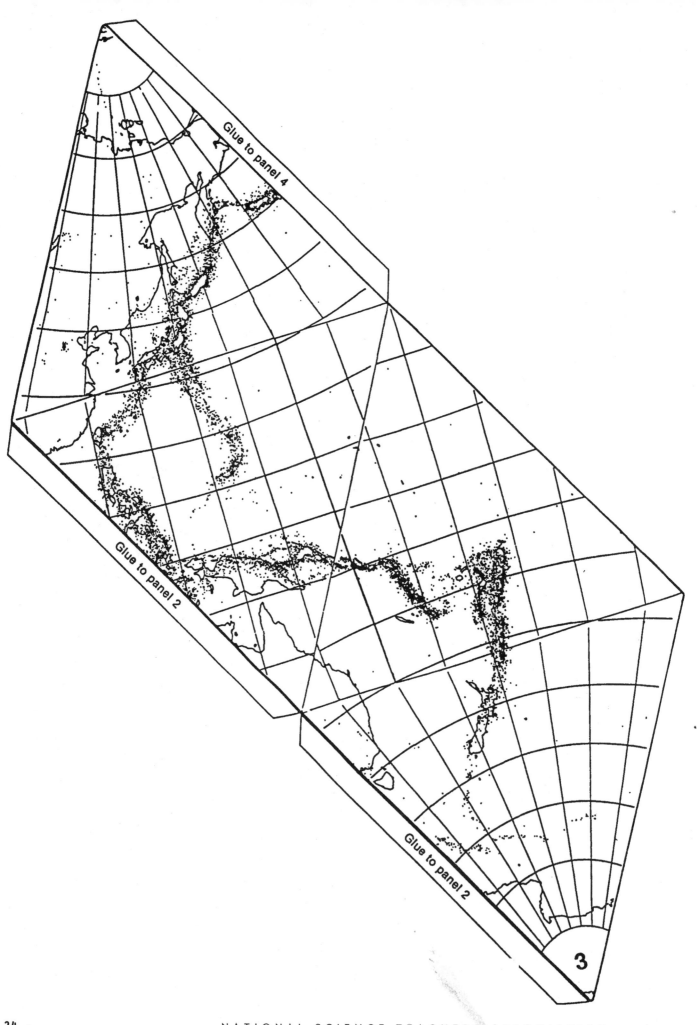

Glue to panel 4

Glue to panel 2

Glue to panel 2

3

NATIONAL SCIENCE TEACHERS ASSOCIATION

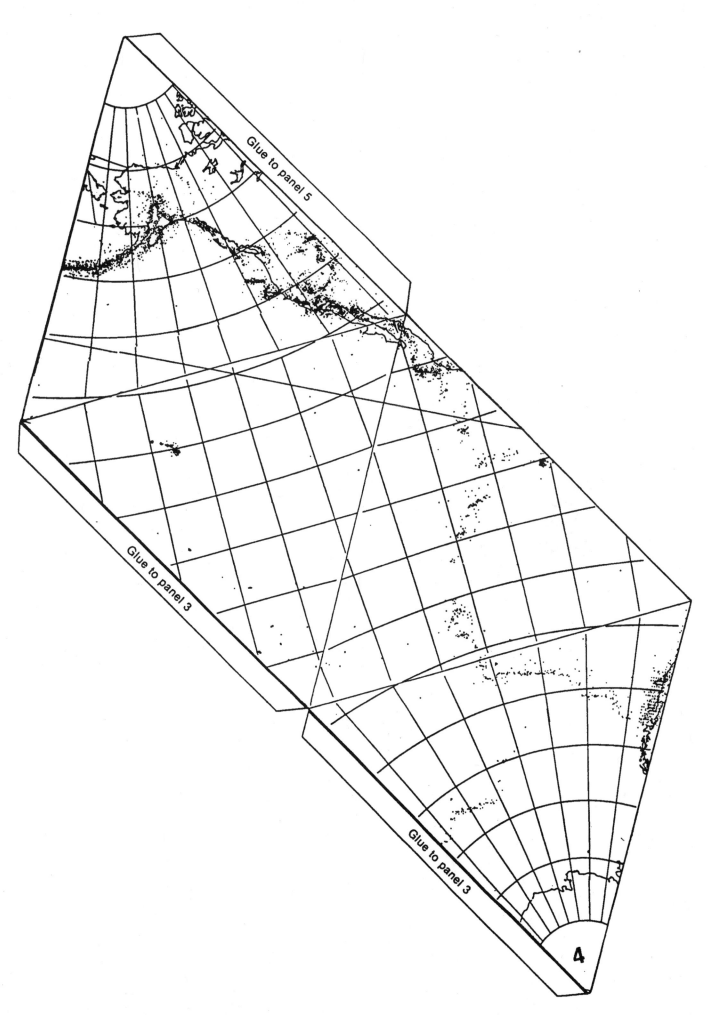

Glue to panel 5

Glue to panel 3

Glue to panel 3

4

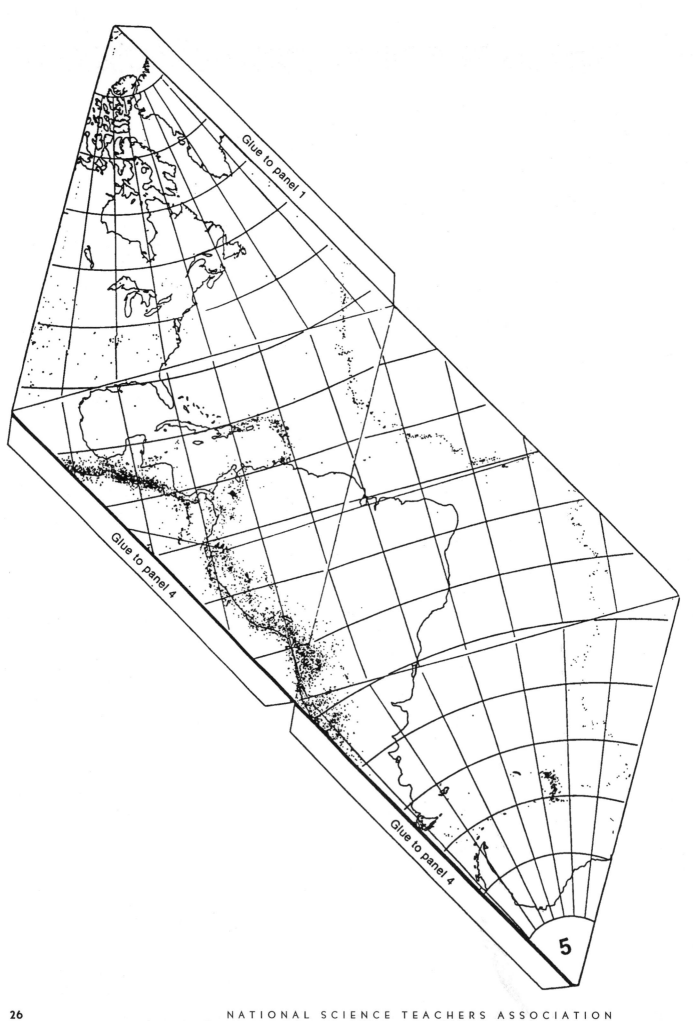

Glue to panel 1

Glue to panel 4

Glue to panel 4

5

GeoPatterns

What Is Happening?

Students need experience with pattern recognition and pattern explanation. They must also learn to distinguish between significant and insignificant patterns in what they observe. In this Activity students look for patterns in earthquake distribution. This project is open-ended; students must decide for themselves whether or not the patterns they identify have significance.

Data on earthquake distribution provide evidence for the theory that Earth's lithosphere (Earth's outer shell) is divided into separate pieces called plates. These plates move over Earth's surface independently of one another, colliding, pulling apart, or sliding past one another. Earthquakes occur when rock at the edges of two or more plates yields to the forces acting upon it. Plate movement releases energy, which is then transmitted to adjacent rock causing it to vibrate or quake.

Studying the distribution of earthquakes gives geologists a precise map of plate boundaries. Certain geologic features—such as ocean trenches and ridges, some volcanoes, and certain types of mountain ranges—occur along plate boundaries too. Geologists study these and other features to learn more about the *type* of boundary at a specific location. Trenches form where two plates come together and one plate slides beneath the other. Volcanoes often form where one plate slides beneath another. A chain of volcanoes forming an ocean ridge can also form at a plate boundary. Mountain ranges may rise at the boundary of two colliding plates, although not all mountain ranges are formed this way.

Encourage students to think about the patterns they observe in earthquake distribution and to extend their study to look for related patterns in the distribution of other geologic features, especially volcanoes.

Although students spend only a short time looking for patterns in earthquake distribution, remind them that geologists have spent decades collecting earthquake data and piecing it together. Help students appreciate the work that went into the data they are analyzing. Recognizing patterns in such data has been crucial in developing the scientific theory of plate tectonics, one of the most basic unifying theories of Earth science.

Materials

Each student or group of students will need
◊ all five panels of the strip map (pages 22-26)
◊ scissors
◊ glue (or clear tape)
◊ map of the ocean floor
◊ colored pencils (optional)
◊ atlas

Important Points for Students to Understand

◊ The occurrence of patterns in nature—such as earthquake distribution—leads scientists to look for explanations.

◊ World earthquake distribution is not random.

◊ Earthquakes and other geologic phenomena—such as volcanoes, trenches, mid-ocean ridges, and mountains—cluster along plate boundaries. The distribution of these phenomena can be used to map these boundaries.

Time Management

Construction of the globe requires approximately 15–20 minutes. The time required for student investigation and problem solving will vary depending on whether students work individually or in small groups. This also can be extended out of class. The central components of this Activity can be completed in one class period.

Preparation

You will need to have on display a map showing the terrain of the oceans. To assist in the placement of data from Table 1 (page 21), have at least one world atlas available in your classroom. Because earthquake and volcanic activity overlap considerably, this Activity can be augmented by presenting maps showing the distribution of active volcanoes on Earth (see the Annotated Bibliography, page 200). Earthquake data can be updated from more recent almanacs; these reference books also contain data on volcanic activity.

When copying the panels for student use, you may wish to use heavy-grade paper, such as card stock, to increase durability.

Suggestions for Further Study

Earthquakes and volcanoes do not *always* occur along plate edges. To appreciate this, students can research two specific earthquakes: the 1811 and 1812 quakes in New Madrid, Missouri, and the 1886 quake in Charleston, South Carolina. Have students investigate why these two earthquakes puzzle scientists, and how geologists explain earthquake activity far from plate boundaries. Students can also research the impact of particular earthquakes on selected communities in the United States and other countries. In a related project, students can explore the science of earthquake prediction, including its methods and limitations. How well can current prediction techniques estimate earthquake location, timing, and intensity? How has plate tectonics theory changed scientists' ability to predict earthquakes?

Have students investigate the differences between the Richter and Mercalli Intensity scales, and learn why each is used to describe earthquake activity. (These earthquake measurement scales are described in Reading 3.)

Suggestions for Interdisciplinary Study

Patterns can be found in any number of data sets. Have students investigate patterns that are found in their own surroundings and ask them to share their findings with the class. Have students design their own data collection tables to record their observations about patterns occurring in their surroundings. Encourage students to appreciate the relationship between accurate and complete data collection techniques when performing science.

Answers to Questions for Students

1. Most earthquakes occur along the boundaries of continents and in mid-ocean, locations that also happen to be plate boundaries. If earthquakes appeared randomly around the world, they would not be found primarily in these regions.

2. Answers will vary depending on student responses to the previous question.

3. Answers will vary. Encourage students to suggest as many *plausible* hypotheses as possible. Ask them how their hypotheses could be tested.

4. Answers will vary. However, the significance of the patterns of earthquake and volcano activity around the world should be made clear to students, and should lead them to the possibility that these two phenomena are related. Encourage students to suggest as many explanations for the relationship between earthquakes and volcanoes as possible.

5. Students should identify various topographic features of the seafloor, such as trenches, ridges, and volcanoes.

6. Students should identify patterns and be encouraged to consider the connections between patterns. Encourage students to suggest as many plausible hypotheses as possible. Ask them how these hypotheses could be tested.

7. Students should notice the relationships among the mid-ocean ridges, oceanic trenches, and earthquake activity. Again, encourage students to suggest as many *plausible* hypotheses as possible. Ask them how these hypotheses can be tested. If concepts related to plate tectonics do not arise in their discussions, suggest them and ask students for their reactions to these concepts.

Orbiter 5 Shows
How Earth Looks from the Moon

There's a woman in the earth, sitting on
her heels. You see her from the back, in three-
quarter profile. She has a flowing pigtail. She's
holding something
in her right hand—some holy jug. Her left arm is thinner,
in a gesture like a dancer. She's the Indian Ocean. Asia is
light swirling up out of her vessel. Her pigtail points to Europe
and her dancer's arm is the Suez Canal. She is a woman
in a square kimono,
bare feet tucked beneath the tip of Africa. Her tail of long hair is
the Arabian Peninsula.

A woman in the earth.

A man in the moon.

May Swenson

All Cracked Up

Background

Many of the objects scientists study, like planets or atoms, are too big or too small to work with by hand. To solve this problem, scientists build **scale models**, like shrinking a planet down to the size of a basketball or blowing an atom up to the size of a baseball. Scale models make objects easier for scientists to study, and they also make it easier to learn about the patterns that can be found in nature.

Good models usually focus on only several parts of the object that is being studied. A model car, for example, shows its shape and appearance but not how the engine works. A classroom globe shows national boundaries but not what kind of rocks make up Earth's surface.

Different kinds of scale models serve different purposes. Geologists sometimes build scale models to study Earth. In this Activity, you will examine several geological models of Earth to learn more about some of Earth's important features.

Objective

To evaluate different models of Earth and to use these models to understand some of Earth's important features.

Materials

Each student or group of students will need
◊ three models of Earth

Vocabulary

Crust: Earth's outermost geological layer.
Mantle: The geological layer directly beneath the crust, composed of solid rock.
Outer Core: The layer beneath the mantle, composed of hot, molten metals.
Inner Core: Earth's central geological layer, composed primarily of solid iron and nickel.

Procedure

1. Your teacher will supply three different objects that can serve as models for Earth. Based on your understanding of Earth, determine each model's strengths and weaknesses. Write your evaluations in Table 1.

	Strengths	Weaknesses
Model 1		
Model 2		
Model 3		

TABLE 1

2. Of the three models of Earth, which do you feel is most appropriate? Why?

One important aspect of Earth that scientists have studied extensively is its interior. Through a variety of techniques, scientists now believe that Earth consists of several geological layers. These layers are all different in their composition and characteristics. Earth's outer layer is the *crust*, and the crust is rigid and brittle, like an eggshell. The crust is composed of a wide variety of chemical elements and compounds. Beneath the crust is the *mantle*, which is composed of materials that are more dense than those of the crust. Although the mantle is composed of solid rock, it is not brittle like the crust. The mantle's characteristics are unique and important in explaining many geological events we observe on Earth's surface.

The outer region of Earth—specifically the crust and upper mantle—is divided into sections called plates. Very slow movement of these plates has been observed, and their movement is the result of forces occurring deep within Earth.

Beneath the mantle are the *outer core* and *inner core*. Scientists have strong evidence to suggest that the outer core is composed of hot, molten (liquid) metals, primarily iron and nickel. Scientists believe the inner core to be solid and, like the outer core, composed primarily of iron and nickel.

3. Now your teacher will conduct a short demonstration using an egg as a model of Earth. During the demonstration, sketch your observations of the egg model in the space below.

4. Based on the demonstration, lable the layers of Earth—
   crust, mantle, inner core, and outer core—in the diagram
   below.

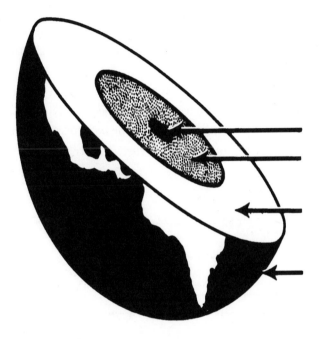

Questions/Conclusions

1. What in the egg model represents each of the four layers?
   What do the cracks represent?

2. Evaluate the cracked egg model for strengths and weaknesses. Place your evaluations in Table 1.

3. How do Earth's layers compare in their composition and
   consistency with the egg layers?

4. Is the cracked egg a useful model for Earth's layers? Why or
   why not?

5. Unlike the eggshell sections, Earth's plates move relative to
   one another. Some plates move apart, others collide. Earthquakes often result from these motions. Predict what other
   consequences might result from plate motion.

6. Why do we use scale models when studying geology?

All Cracked Up

What Is Happening?

Scientific research involves models: scale models to represent objects in manageable sizes, dynamic models to simulate processes, and many others. Students need to appreciate how models are used in scientific research. They also must learn how to evaluate models for strengths and weaknesses. This Activity presents several different Earth models and encourages their close examination and evaluation.

Scientists know Earth's interior has several layers, each different in its composition and basic properties. The outermost layer is the *crust*. Beneath the crust is the *mantle*, and beneath the mantle are the *outer core* and *inner core* (Figure 1, not to scale).

Materials

Each group of students will examine three Earth models. For your demonstration of the cracked egg model, you will need

◊ three Earth models

◊ several hard-boiled eggs, brown or dyed are preferred

◊ one small, sharp kitchen knife

◊ several narrow- and broad-tipped permanent markers

Figure 1

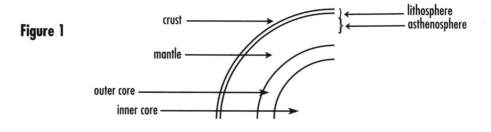

These four basic layers of Earth are defined by their composition. The *lithosphere* and *asthenosphere* are defined by their properties, rigid and fluid respectively. The term lithosphere describes a region ranging from Earth's surface to a depth averaging 100 kilometers. The lithosphere consists of the crust and a portion of the upper mantle. It is broken into a series of plates that move independently of one another. The fluid properties of the mantle allow the lithosphere plates to slowly move in response to the forces beneath them. Although slow in human terms—plates usually move at rates of only a few centimeters each year—this motion can be substantial over thousands and millions of years. Interactions of the lithosphere plates can cause earthquakes and volcanoes. They also can result in the creation of mountain ranges, ocean ridges, and ocean trenches.

The region of the mantle directly beneath the lithosphere is the asthenosphere. Students might confuse the lithosphere/asthenosphere sequence with the crust/mantle/outer core/inner core sequence. They are separate. Geologists distinguish between the crust and the mantle on the basis of chemical properties; they distinguish between the lithosphere and the asthenosphere on the basis of physical properties. The material of the asthenosphere exists at a higher temperature and pressure than does the material in the lithosphere. As a result, the asthenosphere can flow. On the other hand, the lithosphere is relatively brittle and can break into pieces. These physical differences are important for understanding plate tectonics and will be covered by later Activities. In this Activity, emphasis need only be placed on the fact that Earth's interior is layered, and these layers possess different chemical and physical properties.

The distance from the surface to the center of Earth is 6370 kilometers. Following are the approximate thicknesses of each layer. Crust: 8 kilometers (under oceans) and up to 70 kilometers under continents. Mantle: 2900 kilometers. Outer Core: 2300 kilometers. Inner Core: 1200 Kilometers (radius).

Important Points for Students to Understand

Earth's interior has several layers, differing in their chemical and physical properties.

Earth's outer layer—the lithosphere—is broken into pieces called *plates*.

Earth's plates move independently of one another. Their movements and interactions produce geologic features and events.

Models are useful in scientific research, but they have both strengths *and* weaknesses.

Time Management

This Activity can be completed in approximately 30 minutes.

Preparation

This Activity is divided into two parts. For the first part, students examine models of Earth for strengths and weaknesses. Present *three* models to students. Possibilities for models include a styrofoam ball, a basketball or tennis ball (with hollow cores), a baseball or golf ball (with solid cores), a regular classroom globe, an apple or orange, or a bowling ball in cross-section. Different clay models can be made as well—this adds flexibility to the modeling process. Do not prompt students with attributes of any of the models selected.

Demonstration

The first portion of this Activity involves student examination of three Earth models. The second portion begins with this teacher-led demonstration.

1. After students have examined the three Earth models, begin the demonstration. Show students a plain, hard-boiled egg and ask them what would happen if you tapped the egg on a hard surface. Discuss with students the characteristics of the shell that cause its brittleness.

2. Gently tap the egg on a hard surface until a pattern of cracks is produced. Using the narrow tip of a permanent marker, outline enough of the cracks so that there are about eight large "plates" and about eight small ones.

3. Discuss with students how the egg's cracked shell provides a useful model for Earth's surface. Introduce the concept of plates in Earth's lithosphere. Ask students to sketch the cracked egg model.

4. Using a sharp, wet knife, cut the egg lengthwise in half. If the knife blade is wet, the egg is less likely to stick to the blade. Also, have another uncracked hard-boiled egg available in case the first one falls apart while cutting. One half of this egg will model Earth's cross section.

5. Make a dot in the center of the yolk cross section with the broad tip of your marker. The dot will represent Earth's inner core. The shell represents Earth's lithosphere—the crust and the rigid part of the upper mantle that together form Earth's plates. If the egg is dyed beforehand, or if you work with a brown egg, the shell's coloring can be identified with Earth's crust. The egg white represents the lower portion of the mantle. The yolk represents the outer core and the colored dot in the yolk's center represents the inner core. Discuss the various layers with your students, and ask them to compare the consistency of Earth's layers with the consistency of the egg's layers.

6. Ask students to identify the strengths and weaknesses of the cracked egg model. Be sure they understand that one major weakness of the model is the fact that Earth's plates can

move independently of one another. Also point out that the movement of the plates can cause geological phenomena, such as earthquakes, to occur on Earth's surface.

Suggestions for Further Study

Different scientific models show different things and different dynamic processes. Sometimes a geological model, for example, will lose significant topographical features, such as mountain ranges and river valleys, because of the need to use a much smaller scale in order to study Earth's dynamism. Students may have difficulty appreciating how reductions in Earth's scale can flatten mountains and valleys, or make all but the largest lakes and rivers disappear.

Have students construct their own scale models of Earth's layers in cross section by using clay or other materials. Encourage students to be creative (but accurate) in their choice of materials and model design. Have them pay particular attention to what happens to certain topographical features when different scales are used. They can then compare and contrast their findings with pictures of the topographical features with which they have been working.

To give students a better understanding of how relatively insignificant changes over a short time can amount to something significant over geologic time, have students calculate the consequences of 1 centimeter of movement per year for 1 million years, 10 million years, and so on.

Suggestions for Interdisciplinary Study

Have students investigate the use of models outside geology. Topographic maps are models of landscapes. The solar system is used to model atoms—electrons revolve around the nucleus like planets revolve around the sun. Computer modeling and simulations are now widespread, and students can explore these applications. Models are common outside the sciences too. Ask students to compare the use of models in science with the use of metaphors and analogies in literature, films, and music. What are "role models" and how do they relate to what is done in this Activity?

Ask students to locate descriptions of Earth developed in other cultures and in other historical periods. Students might be interested in learning that few people before Columbus—not many, as they might think—thought Earth was flat. Ask them to investigate how Earth's shape was determined by ancient civilizations.

Answers to Questions for Students

1. From outside to inside the different layers are:

   pigments on the outer surface = crust

   shell and egg white = mantle

   egg yolk = outer core

   dot at center of egg yolk = inner core

   Students also may reply:

   shell = lithosphere (crust and upper mantle)

   The cracks represent the boundaries between the plates that comprise Earth's lithosphere.

2. Students should identify as a weakness the fact that "plates" on the egg are stationary while Earth's plates move.

3. Answers should include some of the following points: the outer portion of both Earth and the egg are solid and brittle, while the underlying layer is solid but not brittle; Earth's outer core is fluid and the inner core is rigid, and the egg has only one true "core," which is rigid. Of course, the materials that make up Earth and the egg are very different.

4. Answers will vary, but should include the fact that both Earth and the egg are made up of layers with different characteristics. On the other hand, the egg's shape is not a good representation of Earth.

5. Possibilities include mountain ranges, volcanoes, and rift valleys.

6. While student answers will vary, they should be made aware that models are used in many instances to study subjects where the real thing cannot easily be studied. Models can significantly enhance learning, but they also have important limitations.

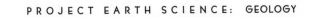

Goodnight

An oyster that went to bed x-million years ago,
tucked itself into a sand-bottom, yawned (so to speak),
and woke a mile high in the Grand Canyon of the Colorado.

John Ciardi

Edible Tectonics

Background

Plate tectonics is one of geology's central theories. At once, it explains a wide variety of observations and phenomena. It explains, for example, the distribution of earthquakes and volcanoes throughout the world. It also explains how many of Earth's surface features—such as mountain ranges, ocean trenches, and fault lines—were formed.

To understand plate tectonics remember that the lithosphere is broken up into a number of plates. Some of these plates are large, while others are small. Although the plates touch they are not connected to each other and move independently. Investigating plate movements—where they move and what causes them to move—is what geologists who are interested in plate tectonics do.

Several hypotheses have been offered to explain plate motion. Because plate tectonic hypotheses are difficult to test, not all geologists agree on which hypothesis is best. One popular hypothesis emphasizes the different characteristics of Earth's layers, and the different ways those layers behave. Many geologists think the asthenosphere is made of solid rock, but the extreme heat and pressure cause the solid rock to flow. These geologists think rock in the asthenosphere flows about two or three centimeters every year. Because the lithosphere rides on top of the asthenosphere, many geologists think flowing rock within the asthenosphere causes plate motion in the lithosphere.

Plate tectonics is a unifying theory that helps geologists explain many of Earth's geological processes and physical features. In areas where the lithosphere plates move apart, for example, rift valleys along the crests of mid-ocean ridges can form. Mountain ranges can form in areas where the lithosphere plates move together and collide. In areas where one plate slides beneath another plate after they collide, ocean trenches and volcanoes can form. This Activity uses a scale model to introduce some of these basic concepts in plate tectonics.

Objective

To investigate how plates move about on Earth's surface and to observe how geologic features form as a result.

Materials

Each student will need
◊ one small Milky Way™ candy bar
◊ towels for clean up

Vocabulary

Plate tectonics: The theory and study of plate formation, movement, interaction, and destruction.

Procedure

1. Obtain a small Milky Way™ candy bar and a paper towel from your teacher.

2. Carefully unwrap the candy bar and use your fingernail to

make a few cracks across the middle portion of its top. The cracked chocolate models the plates of Earth's lithosphere.

3. Hold the candy bar top facing up, with your left thumb and forefinger holding the sides of one end and your right thumb and forefinger holding the sides of the other end.

4. Slowly stretch the candy bar, pulling it apart a few centimeters at most. The chocolate should separate, exposing the caramel. The exposed caramel represents new material that can rise to Earth's surface.

5. Slowly push the stretched candy bar back together again. The brittle chocolate may crumble. On the other hand, "mountain ranges" may form when pieces of chocolate "plates" collide. Alternatively, one chocolate "plate" may slide beneath another.

6. Continue to slowly pull the candy bar apart and push it back together again. Do this until you have a good sense of how plates can be moved about by the motion of the caramel underneath. When the plates are pulled apart material from beneath can move to the surface. When plates are pushed together they can collide, or one can slide beneath another.

7. Once you have finished, pull the candy bar completely apart. Look at its exposed interior and think of the candy bar as a model of Earth's layers. The top layer of chocolate represents Earth's brittle lithosphere, broken into plates. The caramel and nougat represent the asthenosphere, where the material is solid yet still able to flow (Figure 1).

Figure 1

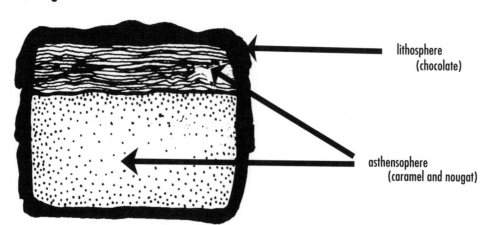

lithosphere
(chocolate)

asthensophere
(caramel and nougat)

8. After answering the questions below, dispose of your model as instructed by your teacher. Be sure to clean up and to wash your hands.

Questions/Conclusions

1. Describe the consistency of the candy bar layers. How do they compare and contrast with one another?

2. Using the candy bar as a model for a portion of Earth, what do each of the candy bar layers represent?

3. Describe what you observed when the candy bar was pulled apart. What might you expect to see at a point on Earth where two plates are moving apart?

4. Describe what you observed when the stretched candy bar was pushed together. What might you expect to see at a point on Earth where two plates collide?

5. From your study of plate tectonics, explain the frequent occurrence of earthquakes along the boundaries between plates.

6. One limitation of this model is that human effort—your fingers pulling and pushing—and not natural geologic processes causes "plate" motion. What natural processes might cause the motion of Earth's plates?

Edible Tectonics

Materials

Each student will need
- ◊ one small Milky Way™ candy bar
- ◊ towels for clean up

What Is Happening?

Middle-level students sometimes have difficulty understanding plate tectonics. The idea of entire continents moving around over Earth's surface can be a hard one to accept—it was for many scientists in the past! This simple demonstration provides a visual and physical reference for students being introduced to plate tectonics.

As will be discussed even more thoroughly in later Activities, Earth's crust is a relatively brittle layer called the lithosphere. Beneath the lithosphere is a portion of the mantle—the asthenosphere—in which solid rock actually flows, although this flow is relatively slow, usually about two or three centimeters each year.

In this Activity, a Milky Way™ candy bar models the relationship between the lithosphere and the asthenosphere. The outer top layer of the candy bar—the brittle chocolate—represents the lithosphere. The next layer, caramel, represents the flowing material in the upper asthenosphere. The layer below the caramel, or nougat, represents the lower asthenosphere where flow also occurs, but not as readily as in the upper asthenosphere.

A drawback of this model is that the continents, which ride atop the lithosphere plates, are not represented among the candy bar's layers. This is significant because it is the continental materials that rise up to form mountain ranges when lithosphere plates collide, while one of the lithosphere plates slides beneath the other to be remelted in the asthenosphere. However, the model *does* convey the interaction between the asthenosphere and the lithosphere, and it is this relationship that is important for students to grasp at this stage in their introduction to geologic concepts. In-depth investigations of the relationships between specific geologic features—such as rift valleys, mid-ocean ridges, trenches, and volcanoes—occur in later Activities.

Students might want to know why testing hypotheses is so difficult and why geologists don't all agree. A good question, but there is no easy answer. But, it might provide an opportunity to discuss the way geologists, among other scientists, actually perform science, and to refresh students' memories about the scientific method. This Activity might also provide an opportunity to discuss how new technologies are enabling today's geologists to test hypotheses in ways that earlier geologists could not.

Important Points for Students to Understand

◊ Earth's interior is layered, and each layer is different in its basic properties.

◊ Earth's lithosphere is relatively brittle and broken into a number of plates, some large and some small.

◊ The asthenosphere is neither liquid nor molten rock. It is solid rock that is under relatively high pressure. The high pressure changes the physical nature of the rock. It becomes somewhat fluid and can flow.

◊ There is a relationship between the movement of the material in the asthenosphere and the movement of the material in the lithosphere.

◊ Certain kinds of plate motion produce characteristic geologic features, such as mountain ranges, rift valleys along the crests of mid-ocean ridges, deep ocean trenches, and volcanoes.

Time Management

This Activity can be completed in about 15 minutes.

Preparation

No special preparations are required. To ensure that the candy bars will perform as expected, go through the Activity prior to the start of class. Also, chill the candy bars enough for the chocolate to be brittle. This Activity may be performed as a demonstration with one large Milky Way™ candy bar.

A note of caution: Candy should be used in classrooms only under close supervision. At the conclusion of this Activity, you might allow students to eat their "models." Be aware of sanitary conditions in making that decision. Because of health reasons (such as diabetes, allergies, and hyperactivity) some students should not consume refined sugar, chocolate, or some of the other ingredients of the candy bar. These students will require especially clear instruction and especially close supervision. Be aware of student dietary restrictions before beginning this Activity to avoid hazards. As a rule, when in doubt about such restrictions, assume a restriction exists. An alternative approach is to demonstrate this Activity while students watch.

Suggestions for Further Study

Have students design other models that will illustrate concepts important to plate tectonics. In designing new models, students must first identify those aspects of plate tectonics critical to understanding the geologic processes involved. Students also should identify strengths and weaknesses of their models.

Suggestions for Interdisciplinary Study

The theory of continental drift was first proposed in the early 1900s by Alfred Wegener (see Reading 1, page 171, for more on the origin of continental drift theory). At the time, Wegener's ideas met with derision and were rejected by a majority of the scientific community. It wasn't until the 1950s and 1960s that mounting evidence—largely provided by technological advances, especially in undersea exploration—led to widespread acceptance of a general theory of plate tectonics, a theory that is considered a direct descendant of Wegener's original hypothesis. Have your students research the development of the theory of plate tectonics, paying particular attention to how accumulated evidence affects a theory's acceptance or rejection. Have your students explore both the good and bad effects of scientists' reluctance to accept theories that seem to challenge generally accepted knowledge.

Answers to Questions for Students

1. The outer layer of chocolate is brittle and can easily be broken into pieces. The caramel and nougat are much less brittle. The caramel and nougat can be reshaped by stretching or compressing, while the chocolate breaks rather than reshaping.

2. The chocolate represents Earth's outer layer, the lithosphere. The caramel and nougat represent the asthenosphere, beneath the lithosphere.

3. When the Milky Way™ candy bar was pulled apart the chocolate fractured into a number of pieces. Where the pieces were not touching, the caramel could be seen beneath the chocolate. Where two plates are being pulled apart a rift valley along the crest of a mid-ocean ridge may form. The rift valley along the crest of the mid-Atlantic ridge is an example of this geologic feature.

4. When the Milky Way™ candy bar was pushed together the chocolate pieces were forced upward and formed ridges, or one piece of chocolate was forced to slide beneath another. Where two plates collide mountain ranges may be formed. The Himalayas were formed when the plates carrying Asia and India collided. At other locations, where continents are not present, one plate sliding beneath another may result in an oceanic trench (extremely deep water). Such areas are called *subduction zones*. The Marianas Trench in the western Pacific Ocean is an example of this geologic feature. Volcanoes are also common subduction zone features.

5. At plate boundaries tectonic forces cause plates to move together, to move apart, or to slide alongside one another. Rocks near the surface must periodically adjust to this motion, resulting in earthquakes.

6. Current scientific explanations for this will be explored in later Activities. At this point, encourage students to develop as many plausible explanations and models as possible. Ask how each might be tested.

from *Rock*

In the longest time of all come the rock's changes,
Slowest of all rhythms, the pulsations
That raise from the planet's core the mountain
      ranges
And weather them down to sand on the sea-floor.

Kathleen Raine

A Voyage Through Time

Background

The movement of the lithosphere has been going on for at least several hundred million years. Because of this continuous motion the shape and position of land masses—of continents and islands—today is much different from their shape and position in the past. Earth is an ever-changing planet!

Land masses ride like passengers on top of the lithosphere. When two plates collide one usually slides beneath the other (Figure 1). At the same time, land masses riding on the plate that is forced downward usually do not sink with the plate. Instead,

Objective

To model the breakup of the supercontinent Pangaea and chart the subsequent movement of land masses.

plate sliding beneath another plate   continent   newly forming mountains   continent   plate

Figure 1

they push up against the land masses riding on top of the other plate. This is one way that mountain ranges are built. Mountain building in this way is occurring today. India and Asia ride on colliding plates, and the Himalayan Mountains are the result of their collision. Of course, in human terms this process is extremely slow.

About 280 million years ago a number of large land masses came together and formed a **supercontinent** geologists call **Pangaea** (pan-GEE-uh). Pangaea existed for roughly 80 million years, until the lithosphere plates beneath it began to break apart and move away from one another. When the plates moved apart, the land masses riding on top of each plate began to move away from one another also.

Geologists have reconstructed the approximate shape and size of Pangaea by comparing rocks found in today's continents and oceans. Using many different kinds of data, geologists have been able to chart the changing positions and shapes of land masses since Pangaea broke apart.

Plate movement still occurs today, and will continue to occur long into the future. This is important to remember. Although continents may *appear* fixed in position, actually they

Materials

Each student will need
◊ a copy of the three map sheets
◊ colored pencils or crayons: red, orange, yellow, green, blue, purple, tan
◊ scissors
◊ a current world map showing terrain, such as mountains and seafloor (on display)

Vocabulary

Supercontinent: When all or several large land masses are joined together. **Pangaea** was a supercontinent that began to break up about 200 million years ago. It was composed of all the present continents. Supercontinents form and break apart as a result of plate tectonics.

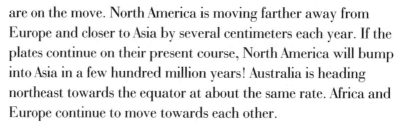

are on the move. North America is moving farther away from Europe and closer to Asia by several centimeters each year. If the plates continue on their present course, North America will bump into Asia in a few hundred million years! Australia is heading northeast towards the equator at about the same rate. Africa and Europe continue to move towards each other.

In this Activity you will follow the movement of land masses over the past 200 million years, beginning with the breakup of Pangaea. A flip-book model will show you how land masses move, and it will reveal to you how our present-day continents arrived at their current positions.

Procedure

1. Your teacher will provide three map sheets, each with a number of frames. These frames are reconstructed maps of the land masses that existed on Earth at a specific time. The interval between successive frames is approximately 10 million years. Frame 20 depicts land masses as they are today.

2. Beginning with frame 20 and working backward, identify the land masses listed in Table 1. Color these land masses as indicated in Table 1. Continue until you can no longer identify the individual land masses.

TABLE 1: COLOR GUIDE FOR LAND MASSES

Land Masses	Color
North and South America	Yellow
Australia	Tan
India	Orange
Africa	Green
Europe and Asia	Red
Antarctica	Blue
Greenland	Purple

3. Beginning with Frame 1 and working forward as far as you can, identify the supercontinent Pangaea. Color this land mass green.

4. Cut out each of the frames along the dotted lines. When all are cut out, stack them in order, 1–20. Frame 1 should be on top.

5. Hold the rectangles along their left side, then flip through the frames. Observe the land masses changing position. You are modeling the breakup of Pangaea and the movement of land masses over 200 million years, arriving at the formation of our present-day continents.

Questions/Conclusions

1. What event began to occur about 200 million years ago?

2. During your coloring of the frames, in which frame did you locate the first appearance of the following land masses:

North America? _____

Australia? _____

India? _____

Europe? _____

Antarctica? _____

3. In which frame did you locate the final breakup of Pangaea? Why did you choose that frame and not another?

4. What causes continents to move across Earth's surface?

5. Sometimes when two plates collide the land masses riding on the plates are pushed together and a mountain range can form. Using a world map, identify two locations where mountain ranges exist and where you hypothesize plate collisions have occurred.

6. How might you explain the fact that continents are *not* submerged when plates collide and one sinks under the other? Explain your reasoning.

7. If mountain ranges can form where plates are colliding, what would you hypothesize might occur where plates are separating? Apply your hypothesis to identify locations on a world map where plates might be separating.

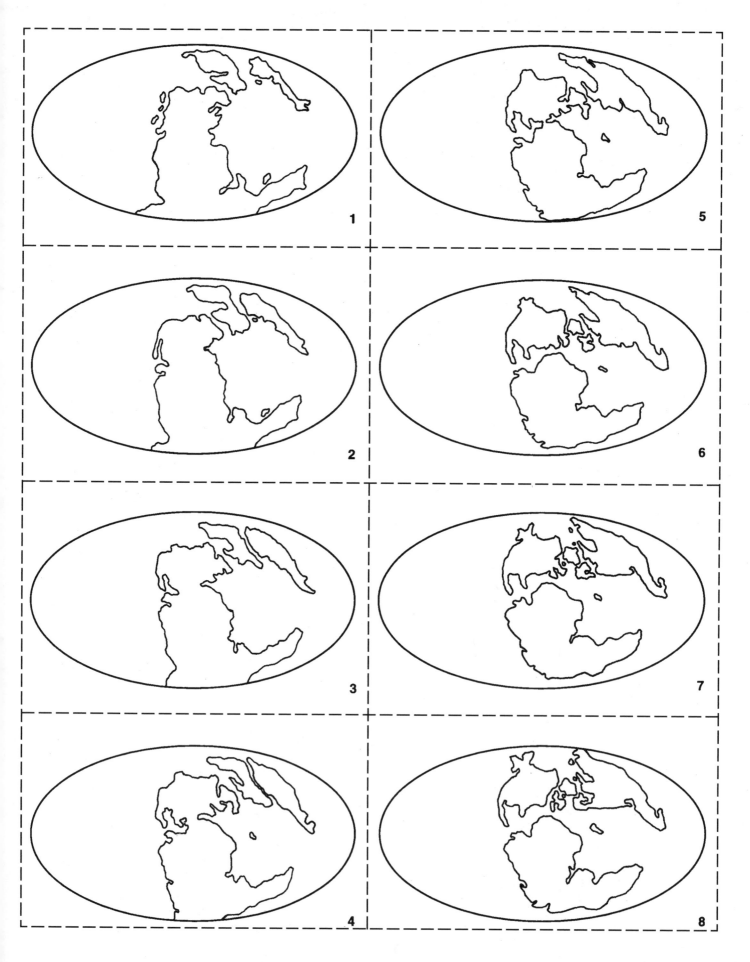

NATIONAL SCIENCE TEACHERS ASSOCIATION

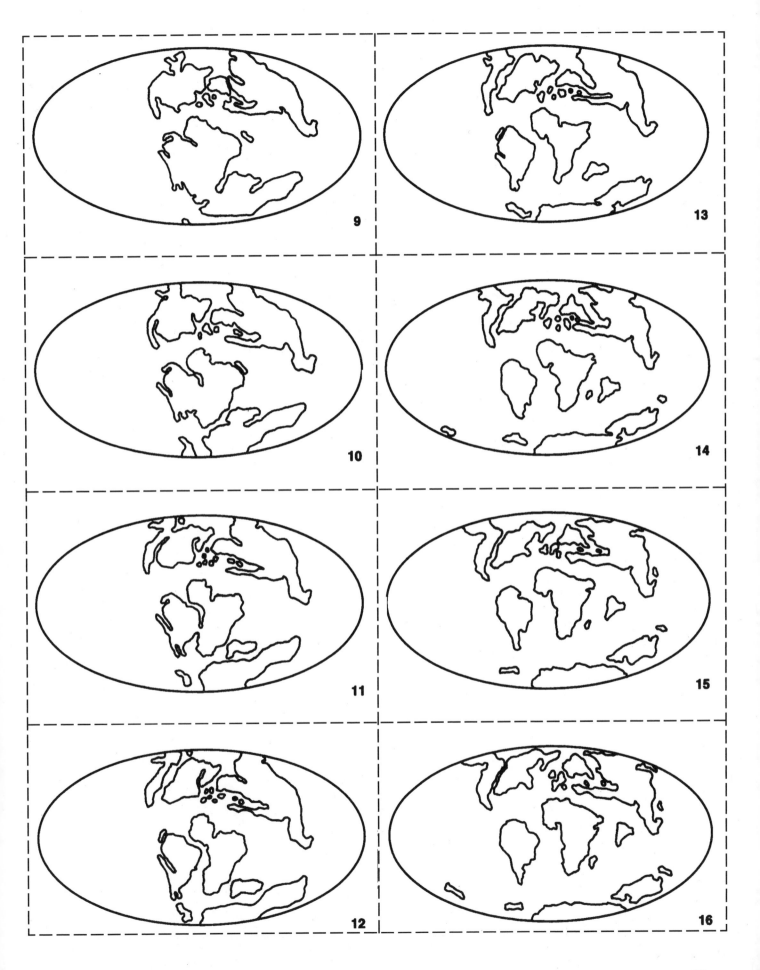

9

13

10

14

11

15

12

16

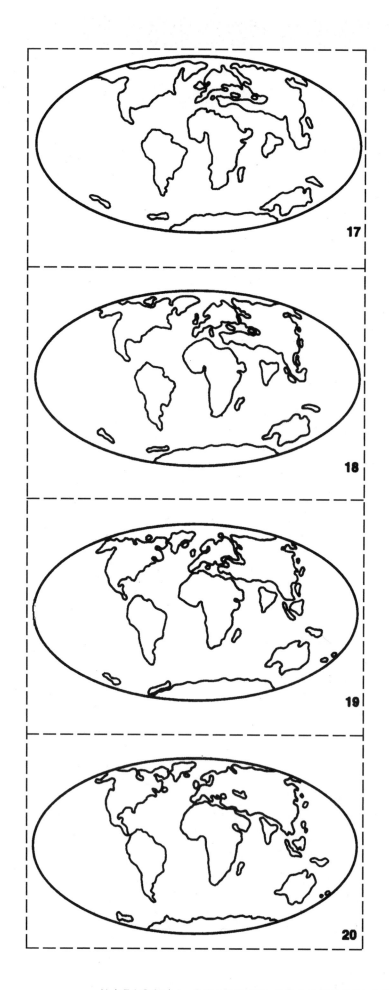

A Voyage Through Time

What Is Happening?

Earth's lithosphere plates may have been in motion since early in its history. Continents "ride" passively atop the plates as the plates move. Several large land masses came together to form one supercontinent approximately 280 million years ago. Geologists call this supercontinent Pangaea (pan-GEE-uh).

Pangaea started to break up roughly 200 million years ago, when the lithosphere plates beneath it began to move apart. Since that breakup, these plates have continued to move, carrying pieces of the former Pangaea with them. The positions we observe today for Earth's land masses are the result of this history of motion. Earth's plates continue to move.

When two plates collide a *subduction zone* is formed, with one plate moving beneath the other. A subduction zone currently exists along the western coast of South America; it is a steep and deep trench in the ocean floor. Because material making up continents is less dense than the material making up the rest of the plates, land masses themselves generally are *not* subducted. Instead, land masses are pushed together at subduction zones. This is one form of mountain building, and it can be observed in the Himalayan Mountains.

Because land masses ride passively on top of plates, the observed motion of continents sometimes is called continental drift. This process continues today. North America is drifting farther away from Europe and closer to Asia by several centimeters each year. If their underlying plates continue on their present course, North America will bump into Asia in a few hundred million years.

Important Points for Students to Understand

◊ Land masses that became the present continents once were part of a supercontinent, which geologists call *Pangaea*.

◊ Land masses have been moving on Earth's surface for at least hundreds of millions of years. They continue to move today at rates of a few centimeters each year.

◊ Land masses are less dense than underlying plates. They remain atop plates even after plates collide and one plate slides beneath another.

Materials

Each student will need
◊ a copy of the three map sheets (pages 52-54)
◊ colored pencils or crayons: red, orange, yellow, green, blue, purple, tan
◊ scissors
◊ a current world map showing terrain such as mountains and seafloor (on display)

Time Management

This Activity can be completed in one class period. Students should be encouraged to discuss their coloring of the land masses with each other, as inevitably they will disagree as to when continents first appeared in their entirety and when Pangaea broke apart completely. Encourage them to discuss the reasoning behind their various decisions. While making these kinds of decisions is largely arbitrary, students should be led toward a clear understanding of the relationship between plate tectonics and continental motion.

Preparation

No special preparation is required for this Activity. To improve the flip book's performance, use heavy weight paper or card stock when reproducing the map frames. Several computer simulations exist that animate the process shown in the flip books (see Appendix B, page 200). Having such a simulation available after this flip-book activity can increase student interest and comprehension.

Suggestions for Further Study

Ask students to determine the location of plate boundaries and the directions of their movement based on different motions observed in the frames. From this students can anticipate future changes in the positions of land masses. Using these predictions, create additional frames for the flip book for the next hundred million years. Could a new Pangaea form at some point in the future?

The motion of land masses has had profound effects on the history of life. For example, after South America and Africa separated (approximately 130 million years ago) primates evolved into two separate lineages: the so–called "old" and "new" world groups. In another example, mammals in Australia evolved in relative isolation and have become biologically unique. Have students investigate differences among animal groups separated by continental movement. The interchange between South and North American biological communities has been investigated in detail by paleontologists. Students may be surprised to learn, for example, that North American opossums are migrants from South America. Ask students to investigate what life on Earth might have been like 200 million years ago when Pangaea started to break apart.

Suggestions for Interdisciplinary Study

Flip books use a basic principle of animation. Have students investigate the use of animation in other areas of science, such as animal motion, biological development through life stages, and geophysical cycles. Have students explore how cartoons are animated to support their investigation.

Answers to Questions for Students

1. The breakup of the supercontinent Pangaea.

2. Answers will vary because identifying a continent's first appearance is largely arbitrary. During discussions of their decisions, students should be encouraged to state their choices and explain the reasoning underlying them.

3. Answers will vary. Encourage students to think in terms of geologic time in their estimations.

4. The movement of the lithosphere plates on which the land masses ride causes continental drift.

5. Possible locations of plate collisions and related mountain building include the Himalayan mountain range (Nepal), the Zagros range (Iran), and the Andean range (South America). Rather than searching for actual locations of plate collision and mountain building, students should use their flip books to identify collision locations and suggest plausible locations for mountain building.

6. Continents are not submerged because the material that composes them is less dense than the material composing their underlying plates. Less dense material rises above more dense material.

7. Answers will vary. A geologic feature often found where land masses are moving apart is a *rift*. The Atlantic-Indian Ocean Rift runs east-west off the southern tip of Africa. The mid-Atlantic Ocean Rift runs north-south down the central Atlantic Ocean. The Red Sea and Great Rift Valley of Eastern Africa is another example.

Earthquake

Still things moving,
        firm become unfirm,
land like ocean waves,
        house like a boat—
a time to be fearful,
        but to delight as well:
no wind, yet the wind–bells
        keep on ringing.

Kokan Shiren

Solid or Liquid?

Background

Think about solids and liquids. A rigid substance can be shattered if it is hit hard enough, and a fluid substance can flow. Solids are rigid and liquids are fluid, right? After this Activity, you may not be so sure.

Scientists use the terms *solid, liquid,* and *gas* to refer to a substance's *phases.* The phases of water are ice (solid), water (liquid), and steam (gas). But not all substances are rigid when they are in their solid phase. Words like *rigid* and *fluid* refer to a substance's qualities, to its characteristics. Other qualities might include hard, flexible, shiny, rough, or many others. All substances have qualities, and all substances take the form of a solid, a liquid, and a gas.

While experience tells us that substances that are in their solid phase are usually rigid, every day we encounter substances that are unusual because they are not rigid when they are in their solid phase. It is very important to be precise when using words like liquid and fluid, or like solid and rigid. They do not always mean the same thing, and *liquid* should be used to define a phase while *fluid* should be used to describe a quality.

In this Activity, you will learn about two more unusual substances—Silly Putty™ and Mystery Substance X—to help you understand one of the main theories about what causes plate tectonics. The asthenosphere is solid rock, yet the solid rock in the asthenosphere is fluid, meaning that it can slowly flow. Geologists hypothesize that a fluid asthenosphere explains plate tectonics.

Objective

To investigate and observe how a substance can, under certain conditions, behave like a solid and, under other conditions, behave like a liquid.

Materials

Part A
◊ Data Tables for each student

Part B
Each group will need
◊ Silly Putty™
◊ hammer
◊ Data Tables for each student
◊ a tray
◊ a board
◊ safety glasses for each student

Part C
Each person will need
◊ Mystery Substance X
◊ towels for cleanup

Procedure

Part A

1. Think about the following typical solids: rocks, books, and shoes. What characteristics do they share? From your group's discussion, complete the section "Typical Solids" in Data Table 1 (page 63).

2. Now, think about the following typical liquids: water, pancake syrup, and motor oil. What characteristics do they share? From your group's discussion, complete the section "Typical Liquids" in Data Table 1.

Part B

3. Send a member from the group to pick up the materials for this Activity.

4. Roll the Silly Putty™ into a ball and bounce it off the table. Pull it, stretch it, mash it. Record your observations on Data Table 2 (page 64). Does the Silly Putty™ display qualities more like a solid or more like a liquid?

5. Roll the Silly Putty™ into a ball and place it on top of your board. Place the board on the floor, in an open area away from your desks and chairs. You are about to test some of Silly Putty's™ other qualities.

6. Before continuing, have everyone in your group (and nearby) put on safety goggles. *Do not continue until this is done*. Select one person to use the hammer. She or he should kneel by the Silly Putty™ ball and wait for the group's signal. The other group members should form observation pairs around *but not directly in front of or behind* the tester.

7. When everyone in the group is safely out of range of the hammer and ready to observe, the group should signal the tester to strike the Silly Putty™. Hit it with one hard swing. What happened to the Silly Putty™? Record your observations on Data Table 2.

8. The tester should peel the Silly Putty™ off the board, roll it back into a ball and repeat the experiment using different amounts of force each time. Before each trial, *be sure everyone is outside the hammer's range.* Record your observations on Data Table 2 after each trial.

9. After several trials, the tester should peel the Silly Putty™ off the board, roll it back into a ball, and place it on a piece of paper. Set the paper and Silly Putty™ in a place where it can rest undisturbed for 30 minutes. Someone in the group should trace a circle on the paper to document the ball's circumference. Record the time.

10. Leave the Silly Putty™ undisturbed for 30 minutes. While waiting, return the other materials to your teacher.

Part C

11. Your teacher will provide you with a small amount of Mystery Substance X. Examine Mystery Substance X— work it in your hands. What are the qualities of Mystery Substance X? When does it have qualities typical of solids? When does it have qualities typical of liquids? On Data Table 2, describe the qualities of Mystery Substance X.

12. Roll Mystery Substance X into a ball and then break the ball apart. Let each piece of the ball rest in one hand. On Data Table 2, describe the qualities of Mystery Substance X.

13. When you're sure you've fully investigated Mystery Substance X's qualities, dispose of it according to your teacher's instructions. Be sure to wash your hands and clean your lab area.

14. Check the time elapsed in the Silly Putty experiment in Part B. After 30 minutes, observe the Silly Putty™. Has its shape changed? Trace a second circle around the Silly Putty™. Handle the Silly Putty™ to learn if other qualities you observed earlier have changed. Record your observations on Data Table 2. Discuss the changes you observe (if any) with members of your group. Is Silly Putty™ a solid or a liquid?

Questions/Conclusions

1. When you bounced, mashed, and struck it with a hammer, did your Silly Putty™ react like a typical solid or like a typical liquid? Explain your answer.

2. When left undisturbed for 30 minutes, what happened to the ball of Silly Putty™? Is this reaction that of a typical solid or a typical liquid? Explain your answer.

3. How is Silly Putty™ different from typical solids, like rocks, books, and shoes?

4. When you worked Mystery Substance X with your hands, what did you observe? Explain why this might have occurred.

5. How is Mystery Substance X different from a typical liquid, such as water, pancake syrup, or motor oil?

6. Apply your understanding of Silly Putty™ and Mystery Substance X to Earth's asthenosphere. How can knowing that solids are not always rigid help you to understand plate tectonics?

7. If Silly Putty™ or Mystery Substance X were used to model Earth's asthenosphere, what would be the strengths and weaknesses of each?

Typical Solid	Characteristics
rocks	
books	
shoes	

Typical Liquid	Characteristics
water	
syrup	
motor oil	

Silly Putty™	Student Observations

Mystery Substance X	Student Observations

Solid or Liquid?

What Is Happening?

Geologists know that the asthenosphere is almost entirely solid rock (there are areas just below the lithosphere where a few percent of the rock is molten). Scientific studies using seismic waves—waves that refract as they move between solids and liquids, and some that stop when a liquid is encountered—provide evidence for this knowledge. Even though the asthenosphere is mostly solid rock, many geologists believe that solid, asthenosphere rock flows, a quality typically associated with the fluid movement of liquids.

Asthenosphere rock flows very slowly and, as it flows, causes overlying lithosphere plates to move in the same direction. Where asthenosphere rock flows to the west, for example, the overlying lithosphere plates move to the west. Where the asthenosphere flows downward into Earth the overlying plates move downward into Earth. Geologists continue to study the relationship between asthenosphere flow and lithosphere plate movement.

But how can solid rock have fluid characteristics; how can it flow? Most of your students probably think of something that's described as being solid as *always* rigid, and of something that's described as being liquid as *always* fluid. If you ask your students for an example of something that's typically solid and of something that's typically liquid, they might even say rock for solid and water for liquid.

This misconception about solids and liquids arises from the mistaken idea that words such as fluid and liquid, or rigid and solid, are interchangeable. In science, as in many other areas, words have very precise meanings, and students must learn to distinguish between the quality of being fluid and the phase of being a liquid. Students must be able to understand the difference between a substance's qualities and its phases. Students must appreciate this important distinction to develop an understanding of plate tectonics.

Qualities describe characteristics the way adjectives do, characteristics such as hard and soft, hot and cold, rigid and fluid, and flexible and stiff. Phases are the physical states that substances assume—solid, liquid, or gas—and are nouns when used to identify a substance as being *a solid*, *a liquid*, or *a gas*. It can be confusing to students, but a substance's qualities are *not* necessarily determined by its phases. A solid isn't automatically

Materials

Part A
◊ Data Tables for
each student (page 63)

Part B
Each group will need
◊ Silly Putty™
◊ a hammer
◊ Data Tables for each student (page 64)
◊ a tray
◊ a board
◊ safety glasses for each student

Part C
◊ one box of cornstarch (estimate about two tablespoons per student)
◊ one 2-liter (or larger) bowl with an airtight lid
◊ one mixing spoon
◊ water
◊ towels for cleanup

rigid, and a liquid isn't automatically fluid. It depends on both the substance and the conditions.

To demonstrate this important point, this Activity has students interact with two substances—Silly Putty™ and Mystery Substance X—to help them learn to be precise in their use of language, and to provide a basis for understanding how a solid rock asthenosphere can flow. This Activity will help students to understand one of the central theories about what causes plate tectonics.

Important Points for Students to Understand

◊ The asthenosphere consists of flowing solids. Many geologists believe this flowing is the driving force behind plate tectonics.

◊ Flow within the asthenosphere is relatively slow, moving at rates between one and ten centimeters per year.

◊ The asthenosphere is neither liquid nor is it molten rock. It is solid rock that slowly flows.

◊ Some solid substances have qualities atypical of solids; some liquid substances have qualities atypical of liquids.

◊ Using precise language is very important in science, as it is in many other disciplines.

Time Management

This Activity, including setup and cleanup, can be completed in one class period or less. However, a 30-minute interval must be provided for Part B of the procedure.

Preparation

Prior to the start of class, assemble the required materials for Part B on trays. After you instruct the class, one member from each group can pick up a tray of materials. This decreases the number of students moving around the classroom and thus decreases the opportunities for an accident.

A note of caution: Silly Putty™ must be struck with considerable force to break or shatter. *Make sure that students wear safety glasses and maintain a safe distance from the testers throughout this Activity.* During the 30-minute waiting period, the Silly Putty's™ capacity to flow can be increased by adding heat. You might suggest that students place the unattended ball of Silly Putty™ on an overhead projector (turned on) or near a heater.

To prepare Mystery Substance X for Part C, mix just enough water with the cornstarch to produce a thick slime. The ratio should be about two or three parts cornstarch to one part water. If the mixture becomes too thin, simply add more cornstarch; add more water if it becomes too thick.

While students are working on Part B, walk around the room stirring the cornstarch slurry with the spoon, casually showing Mystery Substance X to each student group. After students have had ample opportunity to notice the fluid qualities of Mystery Substance X, strike the mixture with your spoon. Students will expect the mixture to splash out of the bowl and will be surprised when it does not. Place a heaping spoonful of Mystery Substance X in each student's hand and ask them to explore its qualities.

Working with Mystery Substance X can be messy. Encourage students to investigate rather than simply play. But also encourage the idea that scientific investigation is fun, while serious. Keep plenty of reusable rags or paper towels on hand for clean up. When your students have finished their investigation, have them return any remaining Mystery Substance X to its original container for reuse and wash their hands. Seal the container with an airtight lid to reduce evaporation.

Suggestions for Further Study

Students enjoy opportunities to experiment with variations of this Activity. Many recipes are available for slimes—substances exhibiting qualities typical of both solids and liquids. Encourage your students also to investigate other substances that mix qualities typical of both solids and liquids.

Place a mound of Mystery Substance X on the counter and put two blocks of wood side by side on top of it. As the cornstarch begins to flow the two blocks of wood will move apart, carried along by the moving cornstarch underneath. Using this demonstration, students can explore how the rigid plates that make up Earth's surface can be carried about by material flowing underneath.

Suggestions for Interdisciplinary Study

Have students read the poem at the beginning of this Activity. Ask them to describe how Shiren's earthquake is similar to or different from what they have observed about solids and liquids. Ask them to describe the mood of the poem. Ask them to think and write about how they feel when substances and objects exhibit unexpected qualities and characteristics, and when events take unexpected turns.

Answers to Questions for Students

1. When bounced, pulled, and stretched Silly Putty™ behaves largely as a typical solid. If left alone it changes its shape without breaking, which is a quality typical of liquids. When hammered Silly Putty™ fractures or breaks, as would a typical solid.

2. When left undisturbed for 30 minutes, Silly Putty™ begins to spread out. The bottom of the ball flattens and looses its roundness. This fluidity is a quality typical of liquids.

3. Typical solids do not exhibit the fluid qualities of Silly Putty™; they cannot be easily stretched.

4. When worked vigorously by hand Mystery Substance X reacts more like a typical solid than a typical liquid. It crumbles, rolls into a ball, and can be broken in half. In this instance, Mystery Substance X reacts like a typical solid.

5. Mystery Substance X exhibits qualities of a typical solid under certain conditions. When worked vigorously it exhibits qualities of a typical solid, but when left alone it exhibits qualities of a typical liquid.

6. Earth's asthenosphere is an atypical solid: it flows. Many geologists believe a flowing asthenosphere causes overlying lithosphere plates to move. Geologists estimate this flow to vary from one to ten centimeters per year, depending on the plate.

7. Answers will vary. Students should recognize that the environment within Earth's asthenosphere is quite different from that of the classroom. Also, flow within the asthenosphere is extremely slow, much slower than either of these substances. Of course, the mantle is not made of either Silly Putty™ or of Mystery Substance X.

from *How the Ground Escapes Me*

How the ground escapes me!
I feel grazing my shoulders
the horizon in flight.

Manuel Altolaguirre

Convection

Background

The lithosphere is divided into at least 13 plates that move independently of one another. A single plate can be as much as 8,000 kilometers across and 90 kilometers thick. The force needed to move such a large object is tremendous!

The plates apparently move because the layer beneath them—the asthenosphere—has a fluid quality even though it is solid rock. The plates ride on top of the asthenosphere and the continents ride on top of the plates. If a portion of the asthenosphere moves in one direction, the plate on top of that portion moves that way also and so do the continents. Sometimes plates move away from one another and sometimes they move toward one another. Sometimes they slide against one another.

But what causes a portion of the solid rock asthenosphere to flow in the first place? In this Activity you will build a model of a **convection cell** in water, experiment with its ability to flow, and investigate it as a possible driving force of plate tectonics.

Procedure

Setup

1. Select a group member to pick up your tray of supplies.

2. Clear off the tray and line it with white paper. This will make observations of your convection cell easier.

3. Place three of the four foam cups upside down on the paper forming a triangle. The fourth cup eventually will be placed right side up amidst the other three, as in the apparatus shown in Figure 1.

4. Add enough room temperature or cooler water to the clear plastic pan so that it is 1/2 to 2/3 full.

5. Place the pan on top of the three upside-down cups.

6. Leave the apparatus undisturbed for several minutes. There should be no ripples in the water when you begin.

Objective

To investigate and observe how material moves within a convection cell.

Materials

For the class
◊ pitchers with room-temperature water placed centrally in the room
◊ a source of hot water, enough to fill three cups for each group
◊ cleaning supplies to handle water spills
◊ food coloring (red, blue, and green)
◊ small containers to hold food coloring
◊ basin for collecting used water

For each group
◊ a tray
◊ a small cup containing food coloring
◊ a clear plastic pan
◊ a pipette or medicine dropper
◊ towels for cleaning the pipette
◊ four foam cups (one with lid)
◊ two sheets of white paper
◊ Data Sheets for each student

Figure 1

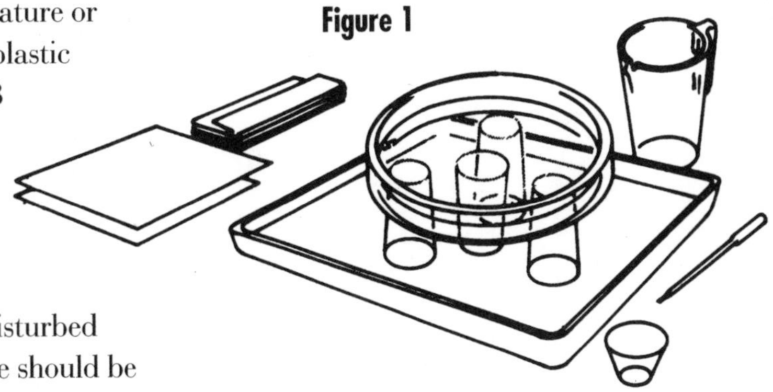

Vocabulary

Convection: The process of heating, rising, cooling, and sinking that many geologists think causes the asthenosphere to flow. A complete cycle is called a **convection cell**.

Trial 1

In Trial 1 you will observe the movement of food coloring with no heat source. You will compare the results of future experiments against this *control experiment*.

7. After the water is still, place a small drop of food coloring at the bottom of the pan in the center. To do this, collect some food coloring in your pipette. Carefully wipe off any excess coloring on the pipette's outside. Move the pipette into and out of the water using slow up-and-down motions only. Place the pipette tip at the bottom of the water at the pan's center (Figure 2). *Take care not to create any movement in the water as you insert and remove the pipette.* Slowly release one very small drop.

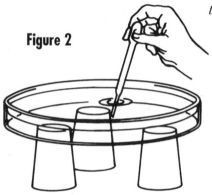

Figure 2

8. Observe the water for about two minutes, viewing both from the top and from the sides. To improve your observations hold a piece of white paper behind the pan. Record your observations on the Data Sheet (page 76). In the space provided draw what you see happening to the distribution of the food coloring. Use arrows to show the direction of movement.

9. After recording the results of this control experiment, gently swirl the water to disperse the food coloring. You only need to replace the water in your pan if it is too dark for further observations.

Trial 2

In Trial 2 you will observe the movement of food coloring when a heat source is placed directly underneath the pan's center, as in Figure 3.

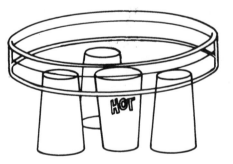

Figure 3

10. After the food coloring from Trial 1 has dispersed, allow the water to become still again.

11. Select one person from the group to retrieve hot water from your teacher in the empty fourth cup. Be certain to use the lid when carrying the hot water back to your table!

12. *Being careful not to disturb the settled water,* gently slide the cup of hot water underneath your pan. Place the cup directly underneath the center of the pan (Figure 3).

13. As in Trial 1, place a small drop of food coloring in the bottom center of the water (Figure 4). Remember to *slowly* release the drop.

14. Observe the water for about two minutes. Record your observations on the Data Sheet. Draw what you see happening to the distribution of the food coloring, using arrows to show the direction of movement.

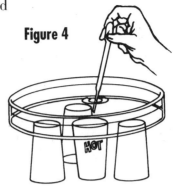

Figure 4

15. After completing Trial 2, remove the cup containing the hot water from underneath the pan and empty it as directed by your teacher. Replace the pan water with clean, room-temperature water.

Trial 3

In Trial 3 you will observe the movement of food coloring with a heat source placed under the center of the pan, as in Trial 2. But in this experiment, the food coloring will be placed on the bottom roughly halfway between the pan's center and its perimeter, as in Figure 5.

16. Allow the water in the pan to become still. As you did in Trial 2, select a person to retrieve hot water from your teacher. Take the empty fourth Styrofoam cup and lid again. *Use caution* when carrying the hot water back to your table! As in Trial 2, gently slide the cup of hot water underneath the pan. Place it directly underneath the pan's center.

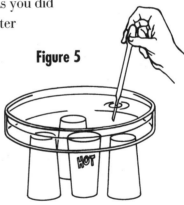

Figure 5

17. Place a small drop of food coloring on the pan's bottom *roughly halfway between the center and the perimeter*. Remember to slowly release the food coloring to avoid disturbing the water.

18. Observe the water for about two minutes. Record your observations on the Data Sheet.

19. When your observations and recording are complete, gently swirl the water to disperse the food coloring. You only need to replace your pan's water if it is too dark for further observations. Remove the cup containing the hot water from underneath the pan and carefully empty it.

Trial 4

In Trial 4 you again will observe the movement of food coloring with the heat source placed under the center of the pan. As in Trial 3, the food coloring will be placed roughly halfway between the pan's center and perimeter. However, instead of being inserted on the pan's bottom, the food coloring now will be placed on the water's *surface*, as in Figure 6.

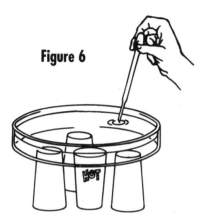

Figure 6

20. Allow the water to settle. As you did in Trials 2 and 3, and select one person from the group to retrieve hot water from your teacher. (Take the empty fourth Styrofoam cup and lid again, and *use caution* when carrying the hot water back to your table!) Gently slide the cup of hot water underneath the pan's center.

21. For this trial, place a small drop of food coloring roughly halfway between the pan's center and edge. But this time, place the drop *directly on the water's surface*.

22. Observe the water for about two minutes. Record your observations on the Data Sheet.

23. After completing your observations and recording, discard the water. Place all experimental materials on the tray and return them to the spot chosen by your teacher. Clean up.

Questions/Conclusions

1. Review the results of the four trials within your group. Contrast the different outcomes to the control experiment. What effect does the heat source have on Trials 2, 3, and 4?

2. For each trial, where in the pan was the current flowing toward the heat source? Where was it flowing away from the heat source?

3. For each trial, where in the pan was the food coloring flowing upward? Where was it flowing downward?

4. This Activity models one of the mechanisms geologists think might drive plate tectonics. In this model, what does the water represent? What does the hot water in the cup represent?

5. In this model nothing represents the plates. To include them, what might be added and where should they be placed?

6. If this model accurately represents a portion of Earth,

explain how plates are moved about on Earth's surface. Why do land masses move when their underlying plates move?

7. The currents in the water cause the food coloring to move at a rate of two to three centimeters or more per minute. Investigate comparisons between this rate and the rate estimated for tectonic plate motion.

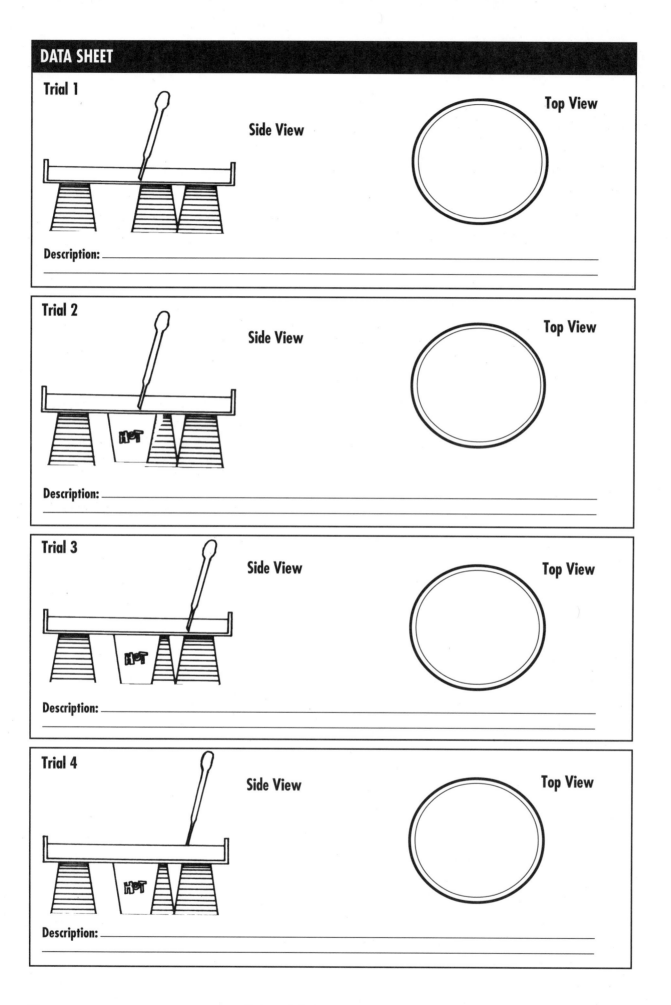

Trial 1

Side View

Top View

Description: _____
_____

Trial 2

Side View

Top View

HOT

Description: _____
_____

Trial 3

Side View

Top View

HOT

Description: _____
_____

Trial 4

Side View

Top View

HOT

Description: _____
_____

Convection

What Is Happening?

Convection is one way heat is transferred from one place to another. Heating a pan of water produces a simple convection cell. When heated from underneath, water nearest the heat source expands, becoming less dense than the cooler water around it. The less dense water rises upward away from the heat source and, as it does so, it begins to cool. By cooling it returns to its original density and sinks. The cycle of heating and rising, cooling and sinking establishes convection currents in the water. The combined system of currents is called a convection cell.

Many geologists believe convection could be the driving force behind plate tectonics. High temperatures and pressures deep within Earth cause the solid rock asthenosphere to flow like a liquid. Heat supplied from Earth's core may cause the asthenosphere to flow in cyclical patterns, such as convection cells. The resulting flow influences the direction of plate motion.

Convection is easily modeled in water. However, take care to distinguish between the *solid* asthenosphere and the *liquid* model used here. Students might think that the asthenosphere is molten, and this should not be reinforced. The asthenosphere is solid rock, but because of the unique combination of heat and pressure deep within Earth, the material in the asthenosphere can slowly flow. (As exampled by the behavior of the substances observed in the last Activity.) When additional heat is supplied from below, convection cells may develop.

Important Points for Students to Understand

◊ Many geologists believe convection plays some role in driving plate tectonics.
◊ The flow of convection cells in the solid rock asthenosphere is very slow, much slower than in the water in this Activity.
◊ The asthenosphere is neither liquid nor molten rock. It is solid rock that exhibits fluid qualities. It can slowly flow.

Time Management

This Activity can be completed in one class period. It also may be divided over two class periods, with observations collected during the first and analyses made during the second.

Materials

For the class
◊ pitchers with room-temperature water placed centrally in the room
◊ a source of hot water, enough to fill three cups for each group
◊ cleaning supplies to handle water spills
◊ food coloring (red, blue, and green are recommended)
◊ small containers to hold food coloring
◊ basin for collecting used water

For each group
◊ a tray
◊ a small cup containing food coloring
◊ a clear plastic pan
◊ a pipette or medicine dropper
◊ towels for cleaning the pipette
◊ four foam cups (one with lid)
◊ two sheets of white paper
◊ Data Sheets for each student (page 76)

Preparation

In this Activity students conduct four experimental trials, each involving a drop of food coloring moved by a convection cell in a pan of water. Observing how food coloring moves in this model allows students to observe how a convection cell functions. Students record their observations and relate what is observed in the pan to what might be happening in Earth's asthenosphere.

Prior to the start of class, assemble the required materials on trays for each group. Use any standard size foam cups; all four cups must be the same size. Clear plastic pans also can vary in size. One convenient source is a garden supply store; drainage trays for plant pots make good pans. Place a few drops of food coloring in small containers. *Warning: food coloring can stain, so caution students to use care to avoid spilling it on clothing, furniture, or the floor.* After instructions have been presented to the class, a group member should pick up a tray of materials. Delegating a single student per group decreases the number of students moving around and reduces the opportunity for an accident.

Because working with hot water can be dangerous, the heating source should be kept under the teacher's close supervision. When students are ready for the hot water, one student from each group should take a foam cup and lid to the hot plate. The teacher should pour the water into the cup, and special care should be taken to minimize the risk of spilling the hot water. Do not fill the cup to the top or students will have trouble carrying the water back to their lab desks. Requiring the use of lids for the cups greatly reduces the risk of spills and burns.

Suggestions for Further Study

Students enjoy opportunities to experiment beyond the scope of this Activity, and should be encouraged to investigate variations of what is presented here. Students can place drops of food coloring elsewhere in the pan to see the results. They can also repeat the trials with water at various temperatures or place the hot water elsewhere beneath the pan. Students should be encouraged to follow a systematic methodology when deciding what additional trials to run, and to compare additional observations against those of previous trials. Substituting rheoscopic fluid (see Novostar Designs listing in Appendix B) for the water and the food coloring is an effective way to observe convection currents; the material that makes the currents visible will not dissipate.

Explore how changes in density play a role in movement within a convection cell. Several density activities in *Project Earth Science: Physical Oceanography* address this concept.

Suggestions for Interdisciplinary Study

Have students investigate other examples of convection and convection cells. Convection drives a variety of phenomena in meteorology, oceanography, and other Earth sciences. Beyond exploring other examples, students should be encouraged to investigate the relative rates of movement within different convection cells. Students can also investigate the role convection plays in situations they might regularly encounter. Heating and air conditioning for homes, schools, and offices usually take advantage of convection cells for circulation. So do stoves, refrigerators, and some fireplaces.

Answers to Questions for Students

1. Students should see that a convection cell is a region of flowing matter. The flow of water in this Activity is in response to the placement of a source of heat under the pan of water.

2. The current was flowing toward the heat source along the bottom of the pan. The current was flowing away from the heat source directly above it, and then along the surface of the water.

3. The food coloring flowed upward in the center of the pan, away from the heat source. It flowed downward along the outer perimeter of the pan.

4. The water represents the asthenosphere, which is actually solid rock. Due to extremes of heat and pressure within Earth the asthenosphere can slowly flow. The hot water in the cup represents the heat sources deeper within Earth's interior.

5. The plates might be represented by a piece of styrofoam or cardboard, and would be found riding on the water's surface.

6. In the convection model, heat from Earth's interior causes convection currents within the asthenosphere. Lithosphere plates move along with the underlying asthenosphere, and thus are carried in the same direction as the asthenosphere flows. Where the asthenosphere flows to the west, for example, the overlying plate moves to the west. If the flow is downward, the plate slides down into the asthenosphere and becomes part of it. If the flow is upward, material from the asthenosphere is added to the lithosphere plates.

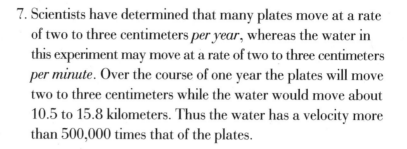

7. Scientists have determined that many plates move at a rate of two to three centimeters *per year*, whereas the water in this experiment may move at a rate of two to three centimeters *per minute*. Over the course of one year the plates will move two to three centimeters while the water would move about 10.5 to 15.8 kilometers. Thus the water has a velocity more than 500,000 times that of the plates.

from *What Happened Here Before*

300,000,000

First a sea: soft sands, muds, and marls
        —loading, compressing, heating, crumpling,
        crushing, recrystallizing, infiltrating,
several times lifted and submerged.
intruding molten granite magma
        deep-cooled and speckling,
                gold quartz fills the cracks—

Gary Snyder

A Drop in the Bucket

Background

The terrain of the **seafloor** is as varied as the terrain of dry land. The seafloor has mountains, valleys, ridges, volcanoes, and many other features. The theory of plate tectonics was developed primarily from information obtained from the seafloor. By studying rocks that make up the seafloor, geologists have learned about how Earth's surface has changed over time. By studying the way trenches and mid-ocean ridges have changed over time, geologists have learned about how Earth's surface also has changed over time.

Geologists take many precise measurements in many different places under the ocean to make sure their seafloor maps are as accurate and complete as possible. Geologists use special techniques, some of which are very expensive. In this Activity you will learn about one of the older, but still useful, techniques geologists use to make seafloor maps. You will also learn about how scientists work in teams, and you will see that how much a technique costs can have a big effect on the whole project.

Procedure

To perform this Activity, the class will be divided into groups of four or more. Each group will have a sector number and, if desired, a vessel name like *Calypso, Explorer, Enterprise, Nautilus, Seafarer, Sounder, Atlantis,* or *Poseidon*.

1. Assign the following roles to group members: sounder, measurer, recorder, and grapher.

2. Select a row of openings in a straight line across the screen or wire mesh. Place a strip of masking tape along the top of the screen next to the row you have selected. Use the permanent marker to number the openings along the tape beginning at one side and crossing to the other. Decide as a team the best way to get the most accurate picture of the features on the bottom of your container using the number of measurements your team can afford (specified by the teacher). Determine the locations along the row of openings on the screen where measurements are to be made.

3. The sounder begins the mapping process by lowering the weighted string through one of the designated openings into the bucket until he or she feels the weight hit the bottom.

Objective

To map and create a profile of a simulated seafloor, and to learn about teamwork and the effects of financial limitations on scientific projects.

Materials

You will work in groups of four or more. Each group will need
◊ one large trash can or bucket (12-15 liters)
◊ one wire screen (hardware cloth) with 1-cm openings
◊ water-soluble finger paint, food coloring, or ink
◊ water
◊ a 25-cm piece of string with a small weight attached to one end
◊ masking tape
◊ one permanent marker
◊ graph paper and pencil
◊ Data Sheets
◊ a meter stick
◊ various non-floating objects, such as rocks, bricks, a bowl, a plaster model of the seafloor

Vocabulary

Seafloor: The bottom of Earth's oceans.

4. The measurer then takes hold of the string where it emerges through the screen (Figure 1).

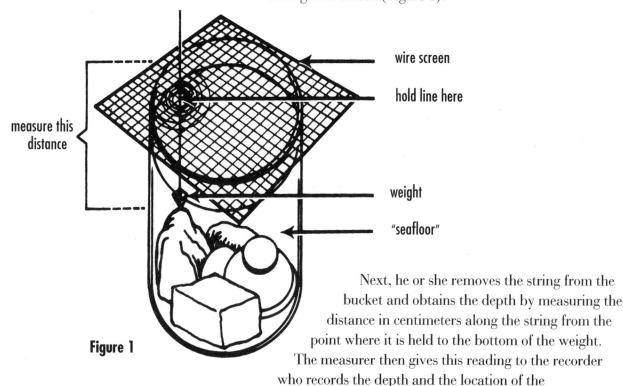

wire screen

hold line here

measure this distance

weight

"seafloor"

Figure 1

Next, he or she removes the string from the bucket and obtains the depth by measuring the distance in centimeters along the string from the point where it is held to the bottom of the weight. The measurer then gives this reading to the recorder who records the depth and the location of the measurement on the Data Sheet (page 86).

5. This process is repeated for all the designated points on the screen. Once all the measurements have been made, the grapher plots the readings on the graph paper and connects the points to create a profile of the bottom.

6. As a class, review the profiles of all the group sector profiles. Next, decide where it would be best to take any additional measurements your group can afford (specified by the teacher). Obtain these measurements using the procedure outlined above and add them to your initial profile. Review all the profiles again to see how and if they have changed.

7. After the profiles are complete, compile all the group sector profiles into one combined profile of the entire mapped area of the "seafloor." The grapher from each group should enter his/her group's sector profile in the appropriate place on the combined class profile.

8. Either remove the water from your container or, if the simulated seafloor is in one piece, remove it from the container. Compare your group's sector profile to the actual "seafloor" to see how accurate it is. Do the same for the combined sector profile and the entire "seafloor."

Questions/Conclusions

1. How does the combined profile compare to the simulated seafloor?

2. How did the accuracy of your group's sector profile change after you plotted additional measurements?

3. If you were to repeat this Activity, how might you improve your mapping technique to get a more accurate profile?

4. In this Activity, you have profiled the bottom of a well-defined enclosed area. Given the varying conditions on the open sea, what are some complications that might be faced by those who attempt to map the real seafloor using this relatively primitive technique?

Group Name:

Sounding Number	Location	Depth (centimeters)

A Drop in the Bucket

What Is Happening?

For many years, scientists believed the seafloor to be an immense expanse of flat, uninteresting terrain. Yet, with new techniques and an intense interest in learning more about the seafloor, geologists have worked hard to map its topography. They uncovered one of the most significant features on Earth's surface—the mid-ocean ridges. This series of mountain ranges extends for more than 65,000 kilometers throughout all the major ocean basins. Knowledge of this and other important geologic features, such as the deep ocean trenches, has provided some of the crucial pieces of evidence that support the theory of plate tectonics.

Special techniques are used to map or profile seafloor features. In addition to sonar soundings to measure depth, special seismic reflection techniques show the thickness of the sediment overlying the bedrock of the oceanic crust. As in any scientific study, the completeness of a seafloor profile depends upon the number, accuracy, and location of the measurements taken. These are limited by time and cost considerations. This Activity demonstrates a crude method of mapping the seafloor, and will help students appreciate some basic considerations that scientists face in working with a project of this type. It also demonstrates the logistics of teamwork and the effects of time and cost limitations on science projects.

Materials

Students should work in groups of four or more. Each group will need
◊ one large trash can or bucket (12-15 liters)
◊ one wire screen (hardware cloth) with 1-cm openings
◊ water-soluble finger paint, food coloring, or ink
◊ water
◊ a 25-cm piece of string with a small weight attached to one end
◊ masking tape
◊ one permanent marker
◊ graph paper and pencil
◊ Data Sheets (page 86)
◊ a meter stick
◊ various non-floating objects , such as rocks, bricks, a bowl, a plaster model of the seafloor

Important Points for Students to Understand

◊ Observations of the geologic features of the seafloor have led to improved theories about processes occurring beneath Earth's surface, specifically the theory of plate tectonics.
◊ Accurate measurement techniques are needed to "map" or profile the seafloor.
◊ Projects such as seafloor mapping involve teamwork and are affected by limitations on time, money, and resources.

Time Management

This Activity can be completed in one class period.

Preparation

Prior to the Activity, prepare a grid on a chalkboard or overhead transparency on which the graphers from each group will graph the combined profile at the end of the Activity (Figure 2). The vertical scale will show depth in centimeters, beginning with zero (the surface) at the top of the axis. The horizontal scale will show measurement location. The grid should be divided into the same number of segments as the number of teams (of four or more members each) in the class. Label the segments with consecutive sector numbers to correspond to the group sector numbers assigned. You can think of the sectors as consecutive regions of the ocean. For instance, Sector 1 might represent a region from the edge of a continent out to a distance of 500 miles offshore, Sector 2 might represent the next 500 miles, and so forth.

Figure 2

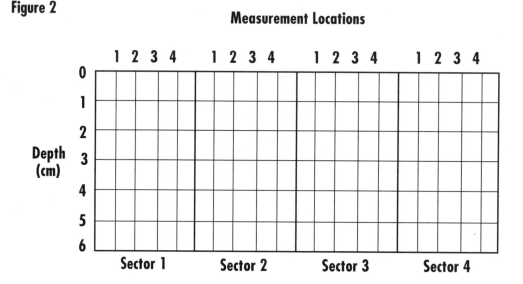

Fill the containers with water, and color the water so that the objects are hidden from view. Water-soluble finger paint works well. Food coloring and ink also work, but care must be taken not to get these coloring agents on clothes.

Place the non-floating objects in the bottom of each trash can or bucket to simulate the seafloor. Make the seafloor sufficiently complicated so that a few measurements are unlikely to provide a complete profile. Anything that does not float will work. Something that can be removed whole from the water and shown to students at the end of the Activity is best. A combination of objects with flat surfaces, like bricks, stacked at different heights will prevent the weight from sliding downhill during measuring.

Cover each container with a piece of wire screen. A total of 15–20 openings across the container works well (Figure 3). Distribute the string and other materials to the groups.

Figure 3

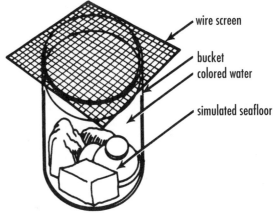

wire screen

bucket
colored water

simulated seafloor

An alternative to this procedure is to use lab sinks in place of the buckets. Simply place the simulated ocean floor on the bottom of the sink, plug the drain, fill the sink, and color the water. Place the screen on the counter over the sink and have the students proceed as directed. This alternative makes it easy to drain the water to study the simulated seafloor and then refill prior to the next class. It is also possible to use a garbage can and pegboard (with sufficiently large holes) rather than a smaller can and wire mesh. The larger container allows for more measurements, and the pegboard provides regular holes for sampling while impeding the students' view of the bottom. Eliminating water altogether might even reduce the mess involved.

Procedure

Begin the Activity by dividing the class into groups of four or more. Assign each group a sector number and a vessel name to correspond with those on the grid, or let them make up their own names. Assign or instruct each group to select its own sounder, measurer, recorder, and grapher.

Tell the groups they can each afford five measurements (depending on the complexity of the "seafloor," three or four might do) from which they must graph its profile. Students must decide where, in a straight line across the screen, to take measurements in order to get the most complete coverage possible. Even spacing over the entire distance generally works best—but don't tell them. Point out that in gathering scientific data, time and money almost always place limits on achieving the ideal. In this Activity, "ideal" would be one measurement per opening, or continuous measurements over the entire profile.

When the groups have completed their sector profiles, collect them and present them to the class for observation. Point out that once some information is available, it can be used to help determine what to do next, such as identifying where to take the next measurement. Tell the groups that they have received additional funding so they can now afford more measurements. Then return the profiles to each group and let them decide where to take additional measurements.

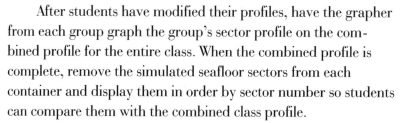

After students have modified their profiles, have the grapher from each group graph the group's sector profile on the combined profile for the entire class. When the combined profile is complete, remove the simulated seafloor sectors from each container and display them in order by sector number so students can compare them with the combined class profile.

Help the groups compare their graphs for individual sectors. Then have the groups compare the combined class profile to the simulated seafloor. Point out the instances in which too few measurements, or measurements not distributed over the width of the screen, have given rise to an inaccurate or incomplete profile. Remind students that the seafloor is actually more varied than the series of flat surfaces used in this Activity, and that much more dramatic depth changes occur in the seafloor's terrain than can be simulated here.

Suggestions for Further Study

Help students create other profiles or multiple profiles of the same "seafloor," which they can then combine to create a more accurate and complete picture. Study topographic maps and profiles of the actual seafloor, and note the major geologic features and dramatic depth changes. Have students look for areas on the maps that correspond to the profiles they have created. Build a scale model of a section of the seafloor with significant geologic features such as trenches and ridges. Have students investigate more advanced methods of seafloor mapping, such as sonar and satellite navigation.

Suggestions for Interdisciplinary Study

Scientific methodology, such as that involved in mapping the seafloor, means collecting accurate, systematic data. Sampling is a technique employed in a variety of endeavors, and is an especially important tool in American politics in the form of public opinion polls. Have students identify the similarities and differences between the sampling techniques scientists use to map the seafloor and those used by pollsters.

Large-scale scientific undertakings require linking effective management skills with effective science. The human genome project is a nationwide, government-backed undertaking that involves identifying and mapping the genes that control all aspects of human life. Have students investigate the management strategies biologists are employing in this investigation, and compare them with those geologists use to map the seafloor. How do they contrast and compare?

Answers to Questions for Students

1. The profile does not include enough detail to show all the features of the seafloor.

2. The seafloor profile became more accurate and included features not disclosed in the first profile.

3. Some possible responses are: take more measurements, be more accurate in taking and recording measurements, and choose different locations for taking measurements.

4. Some possible answers are currents, winds, reefs, drifting, marine life, extreme depth, navigational hazards, rope breaking, and interruptions such as foul weather and rough seas.

from *This Sculptured Earth*

The landscape which we see today is thus
a momentary scene in the stretch of geologic
time. As the world of the past evolved into
that of the present, so the world of today,
which carries traces of its past history,
changes bit by bit into that of tomorrow.
The only constant factor in nature is change.

John Shimer

Seafloor Spreading

Background

About 30 years ago, scientists discovered that there are both age and magnetic patterns in the seafloor. This discovery allowed another piece of the puzzle about plate tectonics to fall into place.

What scientists found was that new seafloor has continually been forming over millions of years at the mid-ocean ridges that wind throughout all Earth's oceans. Molten rock, called **magma**, from inside Earth rises to the seafloor and as it rises it cools and solidifies into new rock. In some places on the seafloor this new rock is pulled apart by the plates' movements, forming two rock masses that move away from each other in opposite directions. Here the seafloor spreads very slowly away from the ridge. Geologists call this process **seafloor spreading**.

Elsewhere, trenches are formed where one plate slides beneath another. The rock on the descending plate is removed and becomes part of the asthenosphere. Because the older seafloor rock eventually descends into trenches and is removed, the oldest seafloor rocks are only about 180 million years old. The oldest continental rocks, which don't descend into trenches, are as old as 4,000 million years.

In this Activity, you will construct a paper model to investigate the patterns that scientists have discovered in their studies of the seafloor.

Objective

To construct a paper model that will illustrate how the seafloor spreads (is created) at mid-ocean ridges and is consumed (removed) where it descends into trenches.

Materials

Each group will need
◊ one copy of the seafloor spreading model pattern
◊ scissors
◊ crayons or colored pencils (orange, yellow, green, blue)
◊ tape

Vocabulary

Magma: Molten rock, before it reaches Earth's surface.
Seafloor spreading: The process by which molten rock rises, cools, solidifies, and is pulled apart by tectonic plates.

Procedure

1. Cut along the dotted lines of the seafloor-spreading model pattern on page 96.

2. Color the areas indicated on the two strips with crayons or colored pencils.

3. Tape together the orange ends of the strips with the colored sides facing each other.

4. Thread the two strips through Slit B of the larger sheet. Pull one side down through Slit A and the other through Slit C (Figure 1).

5. Pull the strips through the slits so that the same colors on both strips emerge from Slit B and disappear into Slits A and C at the same time.

Figure 1

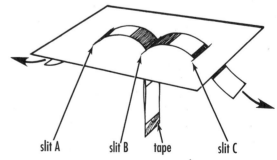

slit A   slit B   tape   slit C

Questions/Conclusions

1. What is happening at Slit B? What feature occurs at the corresponding location on the seafloor?

2. What is happening at Slits A and C? What features occur at corresponding locations on the seafloor?

3. If you were to sample and date the rocks along the colored strip starting at Slit B and moving toward Slit A, what change if any would you see in the age of the rocks?

4. If you were to sample and date the rocks along the colored strip starting at Slit B and moving toward Slit C, what change if any would you see in the age of the rocks?

5. In this model, what do the strips represent? What do the colors represent?

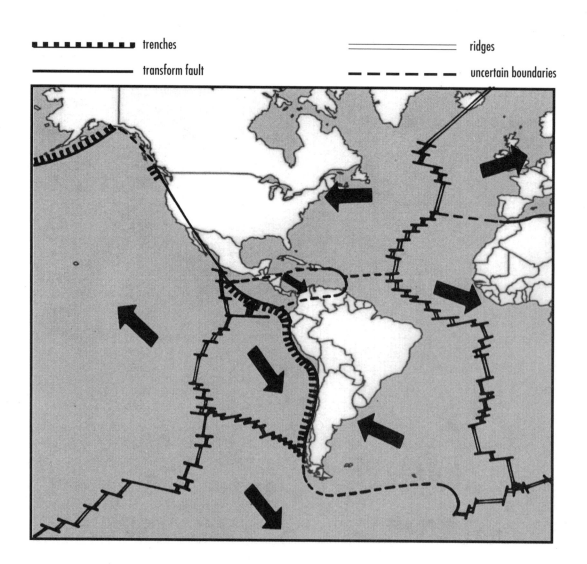

6. New seafloor rock is continually being formed at mid-ocean ridges and old seafloor rock is continually removed at ocean trenches. If the rock on the continents is continually formed but not removed, how would the age of the oldest rocks on the continents compare with the age of the oldest rocks on the seafloor?

7. What are the strengths and weaknesses of this model as a model for demonstrating seafloor spreading?

8. What causes the plates to be pulled apart?

9. Look at the map on these pages showing Earth's plates and the direction of their motion. Which oceans are growing? Which are shrinking? Explain your answers.

World map showing tectonic plates and the direction of their movement

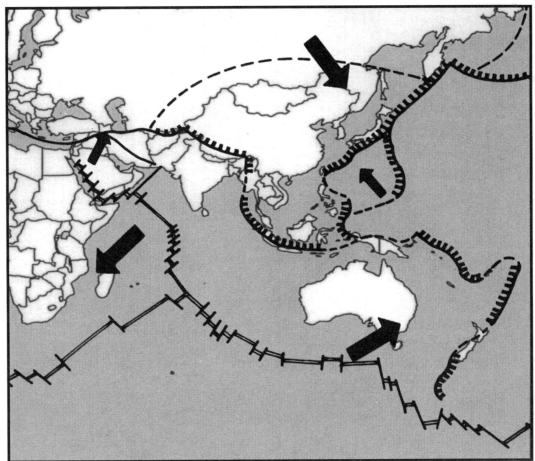

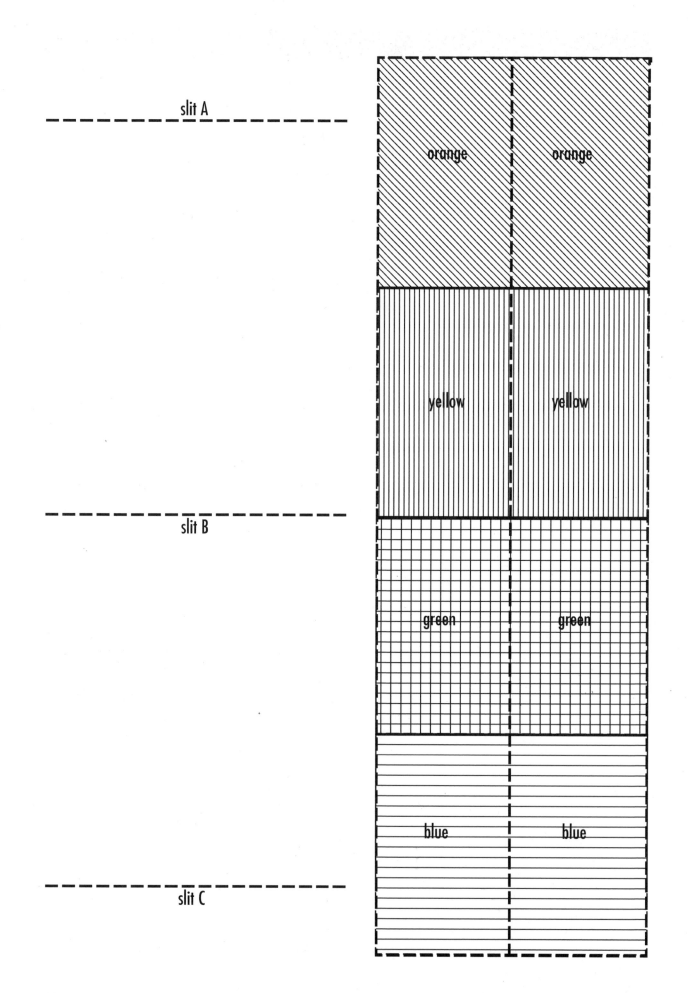

slit A

orange   orange

yellow   yellow

slit B

green   green

blue   blue

slit C

Seafloor Spreading

What Is Happening?

The formation of new seafloor occurs at the boundary between two plates that are separating. As the plates diverge, molten rock from beneath the surface moves upward between them. The molten rock solidifies into a ridge at the surface, is split, and then is pulled apart into two masses by the motion of the plates. The two masses move away from one another in opposite directions. New rock continues to be formed at the ridge, while the older rocks are pushed/pulled away from the ridge on either side. (Reading 1 provides a detailed explanation of this concept.)

Geologists call this movement seafloor spreading, and it results in nearly symmetrical patterns of rock of different ages on both sides of the mid-ocean ridge. The oldest rocks are furthest from the ridge on either side, and the youngest rocks are at the ridge (Figure 2).

Elsewhere on the seafloor, at plate boundaries where one plate overrides another, older rocks are forced down into trenches and are remelted in the asthenosphere. The oldest seafloor rocks are only about 180 million years old because of this cycle. By contrast, the oldest continental rocks (which are less dense and therefore do not descend into trenches) are up to four billion years old.

Rocks with iron in them are weakly magnetic, and scientists have been able to "read" the magnetic alignment of seafloor rocks that have solidified and spread away from the mid-ocean ridge. Not all magnetic seafloor rocks are aligned in exactly the same direction, and geologists have used this fact to support the theory of plate tectonics.

Materials

It is recommended that students work in small groups. Each group will need

◊ one copy of the seafloor spreading model pattern (page 96)

◊ scissors

◊ crayons or colored pencils (orange, yellow, green, blue)

◊ tape

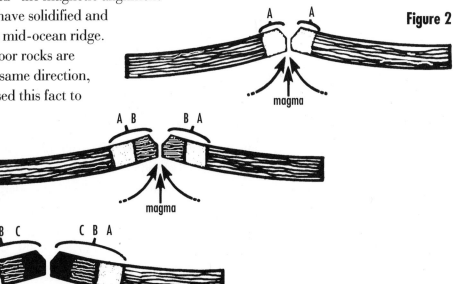

Figure 2

Important Points for Students to Understand

◊ New seafloor is continually created at mid-ocean ridges, which occur where two plates are separating.

◊ The seafloor spreads apart at mid-ocean ridges, creating symmetrical patterns of rock on either side.

◊ The youngest seafloor rocks are found at mid-ocean ridges. Progressively older rocks are located farther away from the ridges on either side.

◊ The oldest rocks are eventually removed at trenches, where one plate slides beneath another.

◊ Continental rock does not descend into trenches; consequently, the oldest continental rock is much older than the oldest seafloor rock.

Time Management

This Activity can be completed in one class period or less.

Preparation

No special preparations are required for this Activity.

Suggestions for Further Study

Have students study a map of the seafloor and locate mid-ocean ridges and trenches, such as the one on pages 94 and 95. Have them create a cross-section diagram of seafloor spreading showing rock formation at mid-ocean ridges and rock consumption at mid-ocean trenches. Have students predict whether relatively old rock or relatively young rock would be found at various points on the seafloor. Also, have them predict the motion of the plates adjacent to mid-ocean ridges and trenches.

Suggestions for Interdisciplinary Study

Have students read the quote that begins this Activity and discuss the meaning behind its final sentence. Have them discuss other aspects of Earth and human life that undergo change. Possibilities include the evolution of living organisms, and of human cultures and societies, the development of individuals and technologies, and many others. Have students select a topic and write a short essay on change, focusing on issues such as: Is change good or bad? How long does it take for change to occur? How do we know that change has taken place?

Answers to Questions for Students

1. Formation of new seafloor. A mid-ocean ridge.

2. Consumption or removal of old seafloor. Ocean trenches.

3. The age of the rocks would increase.

4. The age of the rocks would increase.

5. The strips represent the plates that make up the seafloor. The colors represent rocks of different ages.

6. The oldest continental rocks would be older than the oldest rocks on the seafloor.

7. Answers will vary, but encourage students to evaluate as many aspects of this model as possible.

8. Earth's plates move as a result of movement of material within convection currents in the asthenosphere.

9. The Atlantic Ocean is growing along the mid-ocean ridge that runs its length. The Pacific Ocean is neither growing nor shrinking. Trenches located near South America, Alaska, and eastern Asia are continually removing seafloor as new seafloor is being added along the Pacific's mid-ocean ridges.

Volcano
Tall grey cone
Strong and powerful
Silent while sleeping
Mountain

A Japanese Syntu

Volcanoes and Plates

Background

The lithosphere is made up of tectonic plates that move independently of one another. Some of the plates are moving together, some are moving apart, and still others are sliding past each other. The zone, often thousands of kilometers long, where plates meet when they are moving together is called a **convergent boundary** (Figure 1).

Objective

To investigate the relationship between the type of volcanic activity and related plate boundaries.

Materials

◊ Data Sheet
◊ World Grid Map
◊ World Map with Plate Boundaries
◊ colored pencils: red, blue, and green

Figure 1

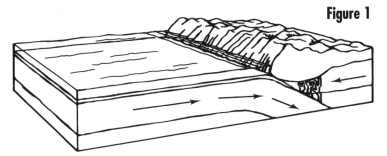

The zone where plates are moving away from each other is called a **divergent boundary** (Figure 2). This is a divergent boundary that has resulted in a rift valley (i.e. crest of some ocean ridges and Great Rift Valley of East Africa).

Figure 2

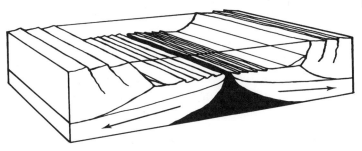

The zone where plates are sliding past each other is called a **transform boundary** (Figure 3). The San Andreas Fault in California is a famous example of a transform boundary.

Figure 3

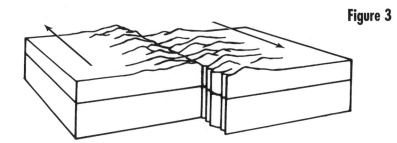

Of the three types of plate boundaries, volcanoes occur most often over the first two, convergent and divergent boundaries. (Volcanoes also occur over hot spots, which will be covered in detail in Activity 10.) Because the plate boundaries are so different—moving toward each other and moving away from each other—the volcanoes that occur in each zone are different.

Volcanic rocks found in the two zones also differ. Andesite is a dark grey rock, and rhyolite is a grey or pink rock. Andesite and rhyolite are both found where plates converge. Basalt is a black rock found where plates diverge. Color differences result from the different amounts of a substance called silicon dioxide that are found in each kind of rock, as well as different amounts of iron and other constituents. Geologists look for clues in volcanic rocks to tell them what type of plate boundary exists at that location or that existed at a given location in the past.

Vocabulary

Convergent boundary: The zone between two plates that are moving together.
Divergent boundary: The zone between two plates that are moving apart.
Transform boundary: The zone between two plates that are sliding past each other. Also called a lateral fault.

Procedure

In this Activity, you will plot the locations of volcanoes using a different colored pencil for each rock type associated with it. From your finished map you will determine the type of plate boundary that occurs at the locations you have plotted.

1. Using the longitude and latitude coordinates on the Data Sheet (page 104), plot all volcanoes on the World Grid Map (page 105). (Note: this is only a partial list of the world's active volcanoes.) The Data Sheet also indicates the percent of several substances found in magma at each location. These substances are silicon dioxide (SiO_2), aluminum oxide (Al_2O_3), and two compounds of iron oxide ($FeO + Fe_2O_3$). Variation in the relative amounts of these substances results in different types of rocks. Use a different colored pencil for the rock type associated with each volcano as follows:

andesite	red
rhyolite	blue
basalt	green

2. Sketch in lines where you think plate boundaries occur. On what data did you base your location of the plate boundaries?

3. Write *divergent* where your data indicate two or more plates are moving apart. Write *convergent* where your data indicate two or more plates are moving together.

Questions/Conclusions

1. Based on what you have learned in this Activity, describe the relationship between volcanoes and plate boundaries?

2. What type(s) of volcanic rock is (are) generally found at divergent plate boundaries? What type(s) of volcanic rock is (are) generally found at convergent plate boundaries?

3. Compare your sketch of plate boundaries to the World Map with Plate Boundaries (page 106). How does your map showing volcano locations compare to those on the World Map with Plate Boundaries?

4. Volcanic rocks are often classified based on color, as follows:

rhyolite	grey, pink
andesite	dark grey
basalt	black

   The color of volcanic rocks is partly related to the proportion of silicon dioxide they contain. Which of the rocks listed on the Data Sheet tend to have the highest proportion of silicon dioxide? The next highest? The lowest?

5. Based on the information provided on the Data Sheet, what generally happens to the proportion of iron oxide ($FeO + Fe_2O_3$) in volcanic rock as the level of silicon dioxide (SiO_2) increases?

DATA SHEET

Volcano	Approximate Location		Magma Composition			
	Latitude (degrees)	Longitude (degrees)	SiO_2	Al_2O_3	$FeO+Fe_2O_3$	Rock Type
Pacific U.S.						
1. Lassen, CA	40 N	121 W	57.3	18.3	6.2	Andesite
2. Crater Lake, OR	43 N	122 W	55.1	18.0	7.1	Andesite
3. Mt. Rainier, WA	47 N	122 W	62.2	17.1	5.1	Andesite
4. Mt. Baker, WA	49 N	122 W	57.4	16.6	8.1	Andesite
U.S. Interior						
5. Yellowstone Park, WY	45 N	111 W	75.5	13.3	1.9	Rhyolite
6. Craters of the Moon, ID	43 N	114 W	53.5	14.0	15.2	Andesite
7. San Francisco Peaks, AZ	35 N	112 W	61.2	17.0	5.7	Andesite
Central America/West Indies						
8. Paricutin, Mexico	19 N	102 W	55.1	19.0	7.3	Andesite
9. Popocatepetl, Mexico	19 N	98 W	62.5	16.6	4.9	Andesite
10. Mt. Pelee, Martinique	15 N	61 W	65.0	17.8	4.5	Andesite
11. Santa Maria, Guatemala	15 N	92 W	59.4	19.9	5.9	Andesite
12. Mt. Misery, St. Kitts	17 N	63 W	59.8	18.3	7.3	Andesite
South America						
13. Cotopaxi, Ecuador	1 S	78 W	56.2	15.3	9.7	Andesite
14. Misti, Peru	16 S	71 W	60.1	19.0	5.0	Andesite
Alaska & Aleutian Islands						
15. Katmai, Alaska	58 N	155 W	76.9	12.2	1.4	Rhyolite
16. Adak, Aleutians	52 N	177 W	60.0	17.0	6.9	Andesite
17. Umnak Island, Aleutians	53 N	169 W	52.5	15.1	12.8	Andesite
18. Kamchatka, Russia	57 N	160 E	60.6	16.4	7.9	Andesite
Japan						
19. Mt. Fuji, Honshu	35 N	139 E	49.8	20.6	11.2	Basalt
20. Izu-Hakone, Honshu	35 N	139 E	53.8	14.8	13.0	Andesite
East Indies						
21. Mayon, Philippines	13 N	124 E	53.1	20.0	8.2	Andesite
22. Krakatoa, Java & Sumatra	6 S	105 E	67.3	15.6	4.3	Rhyolite
23. Karkar, New Guinea	5 S	146 E	60.1	16.4	9.6	Andesite
Central Pacific						
24. Mauna Loa, Hawaii	19 N	156 W	49.6	13.2	11.9	Basalt
25. Galapagos Islands	1 S	91 W	48.4	15.4	11.8	Basalt
26. Mariana Islands	16 N	145 E	51.2	17.3	10.9	Basalt
South Pacific						
27. Aukland, New Zealand	38 S	176 E	49.3	15.6	11.9	Basalt
28. Tahiti	18 S	149 W	44.3	14.3	12.4	Basalt
29. Samoa	13 S	172 W	48.4	13.3	12.3	Basalt
North Atlantic						
30. Surtsey, Iceland	63 N	20 W	50.8	13.6	12.5	Basalt
31. Mid-Ocean Ridge	60 N	18 W	48.2	16.5	11.7	Basalt
Africa						
32. Kilimanjaro, Tanzania	3 S	37 E	45.6	10.3	12.6	Basalt

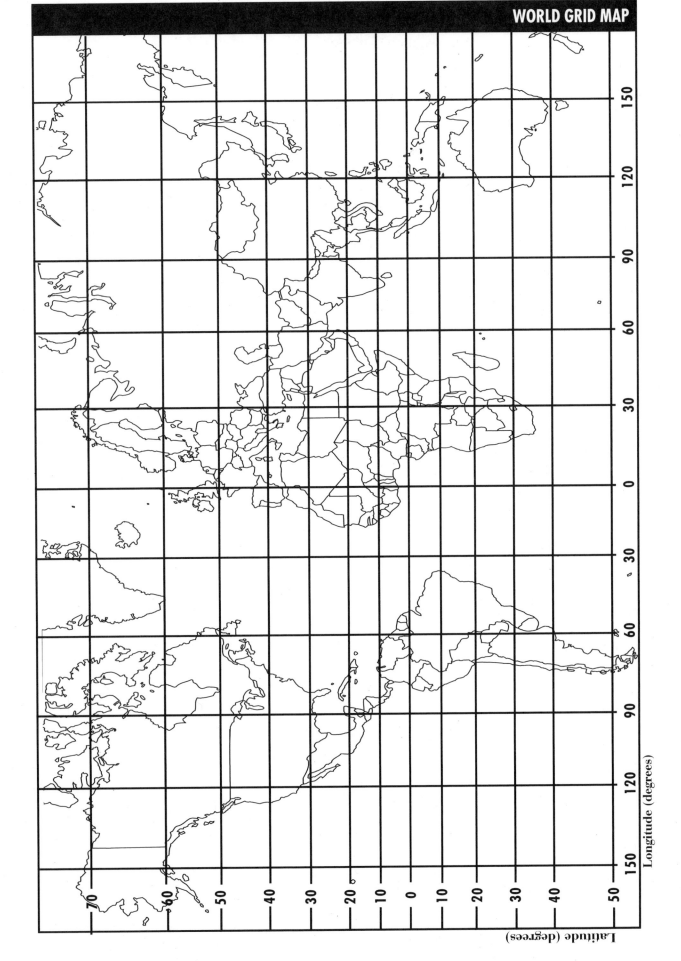

Longitude (degrees)

Latitude (degrees)

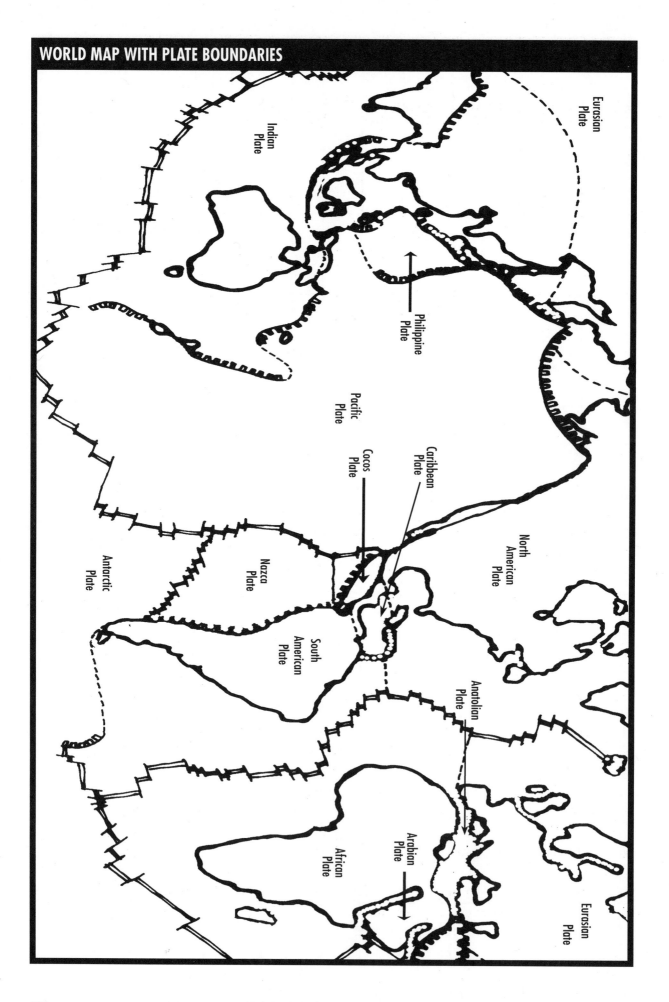

Volcanoes and Plates

What Is Happening?

Plate boundaries are characterized by relatively high levels of volcanic and seismic activity. It was through an extensive study of the pattern of both volcanic and earthquake activity that scientists first began to piece together the theory of plate tectonics. Further study of volcanoes has shown a relationship between a specific type of plate boundary and the volcanoes that occur there. In addition to certain characteristics of the volcanic eruptions, the type of volcanic rock present at a given location depends on the type of plate boundary (divergent or convergent) that occurs at that location. Volcanic rocks at the surface can therefore tell geologists about the processes occurring at depths beneath the location.

Generally, basaltic (black) rock is produced at divergent boundaries as magma from the mantle rises to the surface and solidifies. At convergent boundaries where one plate overrides another, basalt from the sinking plate is remelted forming andesite, a dark grey rock, or rhyolite, a grey or pink rock. Andesite and rhyolite are found at the surface above convergent boundaries, while basalt is found at the surface above divergent boundaries.

In this Activity, students will match volcanoes with rock types on a world map. Once plotted, the information will then be used to deduce the types of plate boundaries occurring at various geographic locations.

A word about the language of plate tectonics is appropriate. The geographic areas where plates converge, diverge, or slide past one another are vast, sometimes covering thousands of kilometers. The geological phenomena that occur within these areas do so over eons. Geological specialists use different terms to describe such phenomena. For example, the word "margin" is often used when describing plate boundaries, as in a divergent margin. The word "zone" is often used to describe the phenomenon of one plate sliding beneath another plate at a convergent boundary—as in a subduction zone. However, to lessen the possibility for confusion among middle-school students the word "boundary" is used consistently throughout the current work to describe the relationships among tectonic plates.

Materials

◊ Data Sheet (page 104)
◊ World Grid Map (page 105)
◊ World Map with Plate Boundaries (page 106)
◊ colored pencils: red, blue, and green
◊ optional, but highly recommended: samples of basalt, rhyolite, and andesite

Important Points for Students to Understand

◊ Plate boundaries are recognized by the frequent occurrence of volcanoes (and earthquakes).

◊ Different types of rock are associated with volcanoes located at different types of plate boundaries.

◊ The type of volcanic rock found in a given area can help determine what type of plate boundary—convergent or divergent—occurs there.

Time Management

This Activity can be completed in one class period.

Preparation

No special preparations are necessary for this Activity.

Suggestions for Further Study

Obtain additional samples of volcanic rock and have students identify them based on their color. Have students research the processes that occur as different types of volcanic rock are formed. Have students create models or diagrams of different types of plate boundaries.

Suggestions for Interdisciplinary Study

A *syntu* is a type of poem that originated in Japan. Its five-line form is governed by a set of rules, which are

Line 1 - One word only

Line 2 - An observation about line 1, using only one sense (sight, touch, etc.)

Line 3 - A feeling, thought, or evaluation about line 1

Line 4 - An observation about line 1 (using a different sense than line 2)

Line 5 - A one-word meaning for line 1

Syntus are typically written about nature. Write the syntu at the beginning of this Activity on the board along with the rules for writing this type of poem and have students compose their own syntus. Have students illustrate and display their syntus.

Answers to Questions for Students

1. Volcanoes occur most frequently on Earth where plate boundaries are located.

2. Basalt. Andesite and rhyolite.

3. Student answers may vary. It is important at this stage to have students engage in the act of comparison in order to understand the relationship between volcanoes and plate boundaries.

4. Rhyolite, andesite, then basalt.

5. It decreases.

Mount Fuji, Opus 5

Flame of fire mountain
Reflecting red in the snow,
Gentle flame, reflecting on the snow's shoulder,
Flame, standing calmly in the sky,
Snuffed in the thick of the night.

Look—there, above it,
Straight above it, among the open spaces on the moon,
In great spirals, winds a blue-green cord.

Drawing near,
I would ask the dragon:

'Why should the ways of the world be sad?
Let the swirling clouds coil no longer,
Let the flames dazzle from your glittering scales...
Lula-lula-la...
Sharp eyes, sharp claws, close them, close them...
Lula-lula-la...See how it coils!'

Kusano Shimpei

Volcanoes and Magma

Background

In 1980 Mount St. Helens in Washington State erupted violently, spewing dust and ash more than 20 kilometers into Earth's atmosphere. The dust from Mount St. Helens traveled around the world. The explosion blew down trees more than 25 kilometers away and rattled windows in houses and office buildings 160 kilometers away. Still, compared with many historic eruptions, such as Vesuvius in Italy and Krakatoa in Indonesia, Mount St. Helens' recent eruption was relatively minor.

Volcanoes that erupt violently are most often located near converging plate boundaries. Where plates converge, a chain of volcanoes forms along the boundary where one plate slides beneath the other. The magma, or lava, associated with this kind of volcano is **viscous**, which means it is very thick, and gases become trapped inside. The gases build up pressure inside the volcano, and eventually explosions occur.

Where plates diverge, most volcanic activity takes the form of relatively minor eruptions or lava flows. This is because magma and lava formed at divergent boundaries is a relatively non-viscous fluid (it flows easily). Volcanoes associated with divergent boundaries are located primarily on the seafloor, but Eldfell in Iceland is an example of a volcano that formed by continual lava eruptions from a divergent boundary. Iceland itself once was a seafloor volcano that grew and grew, eventually breaking the ocean's surface.

But what makes magma rise in the first place? Rocks and minerals melt when subjected to heat. Magma is melted rock that forms below Earth's surface, and it is less dense than the unmelted rock around it. Magma's lower density, along with the action of gases trapped inside it, causes the magma to rise to Earth's surface. In this Activity you will construct a model of an active volcano and observe the forces that cause magma to rise.

Objective

To create a model of a volcano and observe the forces that cause magma to rise to Earth's surface.

Materials

Each volcano model will require
◊ two crayons
◊ one 25-cm piece of string
◊ a beaker, 50-100 ml
◊ a pair of scissors
◊ a paper cup, 300 ml
◊ enough plaster of paris to fill the cup 1/2-1/3 full
◊ hot plate
◊ a spoon
◊ a pan in which to boil water
◊ water
◊ tongs
◊ safety glasses

Vocabulary

Viscous: Very thick; resistant to flow.

Procedure

Part 1 should be completed a day in advance to allow the plaster of paris to harden overnight.

Part 1

1. Remove the paper from the crayons. Break one into pieces and put the pieces in the beaker. Put on the safety glasses and warm the beaker on the hot plate until the crayon melts.

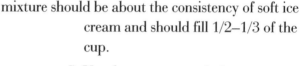

2. Hold the string at one end and use the spoon to push it into the melted wax until it is completely coated. Then remove the string and let it cool. (You may remove your safety glasses when this is done.)

3. Break the other crayon into three or four pieces and bundle the pieces together using the wax-covered string. Leave at least five centimeters of string extending from the bundle.

4. Mix the plaster of paris and water in the paper cup. The mixture should be about the consistency of soft ice cream and should fill 1/2–1/3 of the cup.

5. Use the spoon to push the crayon bundle into the plaster of paris mixture. The bundle should be completely covered and should not be touching the bottom or sides of the cup. Loop the string around a pencil or straw to support the bundle and keep it from sinking to the bottom of the cup (Figure 1).

6. Holding the pencil or straw, tap the cup so that the bundle does not hit the bottom, tap the cup on the table to make any air bubbles rise to the top.

7. Clean your work area and let the plaster of paris harden overnight. To remove wax from the beaker, melt it with hot water, pour out the water, and wipe out the beaker before the wax hardens again.

Figure 1

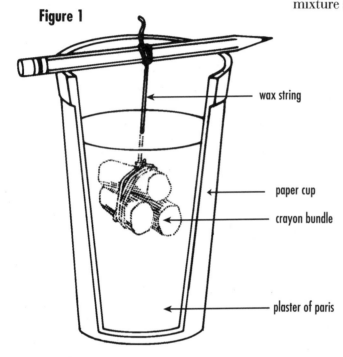

wax string

paper cup

crayon bundle

plaster of paris

Part 2

8. After the plaster has hardened and you are ready to erupt the volcano, tear away the paper cup from around the hardened plaster.

9. Cut off the string close to the surface of the plaster.

10. *Put on your safety glasses and wear them throughout the rest of the Activity.* Use tongs to place the plaster in a pan of boiling water with the string end up. For the best results, the surface of the plaster should be about 1.5 centimeters above the surface of the water (Figure 2).

11. Observe and consider what happens as the wax "magma" inside the plaster "volcano" melts.

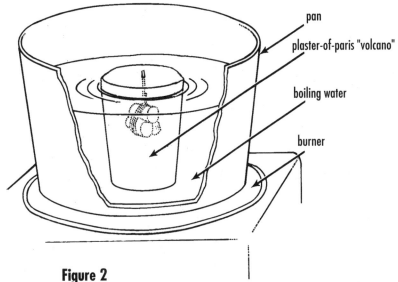

Figure 2

12. When the "eruption" is completed, turn off the burner and allow the water to stop boiling before attempting to remove the volcano. Once the water has stopped boiling, use the tongs to carefully remove the volcano from the pan. Discard the volcano, empty the remaining water from the pan, and clean any remaining wax from the pan, tongs, and burner. *Make sure the burner has had enough time to cool before attempting to clean it.*

Questions/Conclusions

1. Why did the wax from the crayons inside the model volcano rise to the surface when the volcano was placed in the pan of boiling water?

2. What causes magma to rise to Earth's surface in a real volcano? How effective are the wax crayons in the plaster-of-paris model in portraying the action of real magma?

3. When the volcano model was prepared, air bubbles were removed from the plaster. What would have happened if there had been air spaces still in the plaster surrounding the bundle of wax crayons?

4. What might have happened if there had been no waxed string (no opening to the surface) in the plaster to relieve the pressure from the expanding crayon? Is this situation (the absence of a vent for magma and steam pressure) possible in a real volcano? What would be the result?

Volcanoes and Magma

Materials

Each volcano model will require

◊ two crayons

◊ one 25-cm piece of string

◊ a beaker, 50-100 ml

◊ a pair of scissors

◊ a paper cup, 300 ml

◊ enough plaster of paris to fill the cup 1/2-1/3 full

◊ hot plate

◊ a spoon

◊ a pan in which to boil water

◊ water

◊ tongs

◊ safety glasses

What Is Happening?

When rocks and minerals are subjected to certain combinations of heat and pressure, a change in their physical state, such as melting, may occur. Much of Earth's magma is produced at depths of 100-200 kilometers below the surface. Magma is lower in density than the rock that surrounds it, and geologists think this is what may cause it to rise to Earth's surface. The action of rising gases may aid in this process. When magma reaches Earth's surface, it erupts in the form of a volcano. At convergent plate boundaries the eruption is sometimes quite violent. This is because the lava associated with converging zones is highly viscous, or resistant to flow, and the gases trapped inside build up pressure until they are released through an explosion. At divergent plate boundaries the eruption more often takes the form of a continual lava flow.

This Activity uses wax crayons in a plaster of paris volcano to simulate magma as it is heated, expands, and flows to the surface.

Important Points for Students to Understand

◊ Rocks and minerals deep within Earth may melt, forming magma, when subjected to certain combinations of heat and pressure.

◊ Magma is less dense than the rock from which it forms.

◊ Since magma is usually less dense than the rock surrounding it, it tends to rise toward Earth's surface along with certain gases.

◊ When magma reaches Earth's surface it emerges, creating a volcano.

Time Management

This Activity can be completed in two class periods. Part 1 must precede Part 2 to allow the plaster to harden. This Activity may be performed as a demonstration, or the teacher may prepare all the "volcanoes" for the entire class beforehand and allow students to "erupt" them.

Preparation

Collect the materials for the Activity and distribute them to each group. If this Activity is being performed as a demonstration, use Part 1 of the Student Procedure section in preparation for the demonstration, and Part 2 as the demonstration itself.

Procedure

Once all the materials are distributed, have students begin constructing their model volcanoes. *Have students wear safety glasses while melting the crayons.* Make sure that a string is used on all volcanoes. (If no string is used, there will be no pathway for the melted crayon to flow to the surface, and if the pressure becomes high enough the model could explode. Using the string, and making sure the water temperature does not excede boiling, should make this a safe Activity.) Stress to students the importance of ensuring that the plaster contains no air bubbles when left to harden overnight. Air bubbles will interfere with the volcano's eruption by creating spaces in the plaster for the crayon to flow into as it expands.

Before any volcanoes are actually erupted, explain that when the volcano is heated the crayons will melt. This will illustrate that a change in physical state, such as melting, may occur when conditions of temperature and/or pressure change. As the crayons melt, they will expand and rise to the surface along the string.

Explain to students that a similar process occurs deep within Earth when rocks and minerals are heated. They melt, forming magma, which is less dense than the surrounding rock. *When the time comes to "erupt" the volcanoes, make sure students are wearing safety glasses and that they use extreme caution around the hot plates and boiling water.*

Suggestions for Further Study

Volcanic activity can produce tremendous amounts of dust and ash that may become trapped within the atmosphere, leading to a decrease in the amount of sunlight that reaches Earth. Have students investigate the relationship between volcanic activity and climate change. Statistical information about this relationship may be obtained from the U.S. Geological Survey. The 1980 Mt. St. Helens eruption might make an appropriate subject for class study.

Have students investigate some of the peculiarities of volcanic rock, experimenting with mechanisms to explain these occurrences. Here is one such occurrence. When magma reaches Earth's surface, it cools and solidifies into volcanic rock. Some volcanic rock contains older mineral crystals and rock fragments. This means the crystals and fragments either rose to the surface with the magma, or rose through the magma, and then were encased inside the volcanic rock when the magma cooled and solidified. But mineral crystals and rock fragments are more dense than magma. How do they rise to the surface? Magma sometimes contains dissolved gases. Geologists think the action of gases in magma may have helped carry the mineral crystals and rock fragments to Earth's surface, where they were then encased in the newly formed volcanic rock. An easy-to-perform classroom demonstration will help students understand how crystals and rock fragments denser than magma could have risen through it, and will aid in their general comprehension of density. Place several raisins in a beaker filled halfway with a clear carbonated soft drink. Bubbles will adhere to the outside of the raisins and carry them to the surface. There the bubbles will be released into the atmosphere and the raisins will sink, where the process will begin again.

Suggestions for Interdisciplinary Study

Throughout world history, many societies have incorporated volcanoes and volcanic activity within their cultural traditions, especially those societies that have existed near an active volcano. Before the relatively recent advent of scientific explanations, interpretations of volcanic activity assumed a wide variety of forms and were expressed in an equally wide variety of ways. Some societies have continued to reject scientific explanations of volcanic activity, relying instead on tradition and lore indigenous to their particular cultural heritage. Students may want to investigate the different kinds of cultural interpretations among societies that have existed in proximity to volcanoes. There are many interpretations to choose from, from Pompeii to Pinatubo, and student selections might be guided to juxtapose interesting cultural factors, like oral and written traditions.

Such an investigation might be used to introduce students to the idea that modern science is itself a culturally-derived interpretation of physical phenomena. Using volcanic activity as a means of juxtaposing "primitive" and "modern" cultural values might open up other interesting areas of study. Students may want to explore how science and technology have affected indigenous rainforest or island cultures, for example. How have such cultures responded to modern, chiefly Western science, and how does their cultural understanding of science differ from our own? Such an exploration might help students begin to develop an understanding that traditional interpretations of geologic phenomena can tell us valuable and interesting things about the cultures from which they came.

Answers to Questions for Students

1. The crayon expanded as it was heated. The string provided a path to the surface.

2. Rock that is heated to high enough temperatures melts to form magma. This molten rock expands, resulting in material that has a lower density than the surrounding rock. This difference in density between the molten and solid rock causes the molten material to rise toward Earth's surface. Rising gases may also help carry it upward. The wax crayons fairly effectively model this process. However, volcanoes do not have "wicks" for the molten material to follow to the surface.

3. The crayon would have expanded to fill the spaces.

4. The volcano might have exploded as the crayons melted and expanded. Yes, this situation is possible in the case of real volcanoes. Or, the rocks may simply crack, and the cracks then provide passageways for magma to reach Earth's surface.

Born was the island

Born was the island—
It budded, it leafed, it grew, it was green.
The island blossomed on tip, 'twas Hawaii
This Hawaii was an island.
Unstable was the land, tremulous was Hawaii,
Waving freely in the air;
Waved the Earth.
From Akea 'twas fastened together
Quiet by the roots was the island and the land,
It was fast in the air by the right hand of Akea
Fast was Hawaii, by itself—
Hawaii appeared an island.

Traditional Hawaiian Poem

Volcanoes and Hot Spots

Background

Most of Earth's volcanoes form along a plate's perimeter, the boundary where plates are converging or diverging. But volcanoes can also form somewhere inside a plate's perimeter. These volcanoes are caused by concentrations of heat directly beneath a plate. Geologists call these concentrations **hot spots**. Hot spots produce plumes of molten rock, which work their way upward through Earth's crust and form volcanoes (Figure 1). Both the Hawaiian Islands and the Emperor Seamounts in the Pacific Ocean are chains of volcanic islands and underwater volcanoes (many extinct) geologists think formed over a hot spot.

Objective

To study volcanoes formed over hot spots, and to investigate how plate movement is related to the pattern of volcanic island formation.

Vocabulary

Hot spots: concentrations of heat occurring in certain places directly beneath a tectonic plate.

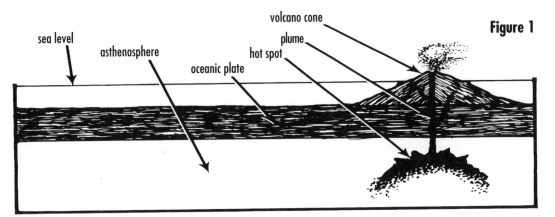

Figure 1

Look at the map of the Pacific Ocean below and locate the Hawaiian Islands and the Emperor Seamounts. Do you see a pattern? In this Activity you will simulate the creation of a volcano over a hot spot and investigate how chains of volcanoes form.

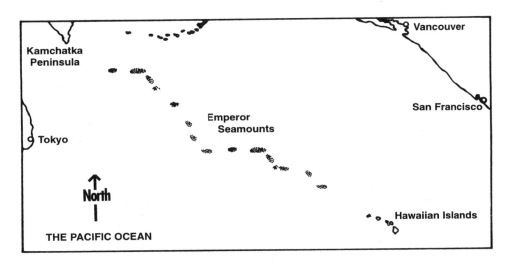

Procedure

Materials

◊ one clear plastic container, such as a shoebox

◊ one small dropping bottle, with small neck opening

◊ red food coloring

◊ hot tap water

◊ cold tap water

◊ a styrofoam "tectonic plate"

To perform this Activity, the class will be divided into groups.

1. Get a tray of materials from your teacher.

2. Fill the shoe box 2/3 full with cold tap water.

3. Add hot tap water to the dropping bottle, then add a few drops of red food coloring.

4. Carefully place the uncapped dropping bottle in the center of the shoe box, making sure that the cold tap water covers the top of the bottle. Do not tilt the bottle when placing it in the cold water.

5. Place the styrofoam "tectonic plate" on the surface of the water so that one end is directly over the bottle (Figure 2).

Figure 2

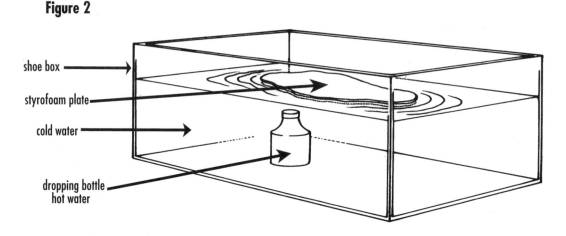

shoe box

styrofoam plate

cold water

dropping bottle
hot water

6. Observe and consider what happens to the hot and cold water. Note where the hot, colored water contacts the "tectonic plate." Sometimes an air bubble can become trapped in the mouth of the bottle. If this occurs, use a pencil, straw, or straightened paper clip to dislodge it.

7. Next, simulate plate motion by gently moving the styrofoam so that a new area of it is directly over the dropping bottle. Do this very carefully so as not to disturb the water in the shoebox. Just touch the top surface of the styrofoam gently with one finger. Observe where the hot water now contacts the "plate."

8. Repeat Step 7 until you run out of room in the box, observing the pattern created among the volcanoes that would form above the plume of rising water.

9. Once you have made your observations, empty out the water from both the shoe box and dropping bottle, and clean your equipment and work area.

Questions/Conclusions

1. What happened to the hot water in the dropping bottle?

2. What happened to the cold water in the shoebox?

3. What is the main reason that hot water (and molten rock) tends to rise?

4. What feature forms on Earth's surface where magma flows upward from beneath a plate?

5. Geologists believe that hot spots remain stationary within the mantle. What then results when a plate moves over a stationary hot spot?

6. Look at Figure 3. In which direction is the plate probably moving? Draw an arrow on the plate to indicate the direction of plate motion.

7. Where is the next volcano in Figure 3 likely to form? Mark the location with an X.

8. Look again at the map of the Pacific Ocean. What pattern do you observe when you look at the Hawaiian Islands? The Emperor Seamounts?

9. The youngest volcano in the Hawaiian Islands is at the southeastern end of the chain. What does this fact tell you about the recent direction of movement of the Pacific Plate?

Figure 3

plate

active volcano

volcano chain

10. Geologists believe that both the Hawaiian Islands and the Emperor Seamounts formed as a result of the same hot spot. If this is the case, what does it tell you about the direction of motion of the Pacific Plate?

11. What are some of the strengths and weaknesses of this model of the way volcanoes form over hot spots?

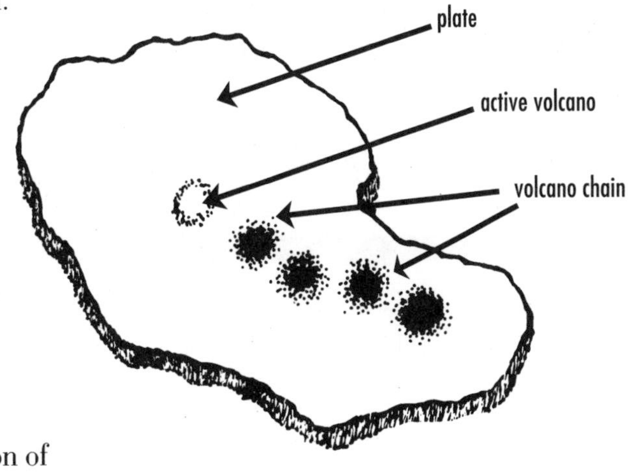

Volcanoes and Hot Spots

Materials

◊ one clear plastic container, such as a shoebox

◊ one small dropping bottle, with small neck opening

◊ red food coloring

◊ hot tap water

◊ cold tap water

◊ a styrofoam "tectonic plate"

What Is Happening?

Tectonic plate boundaries are characterized by the large number of volcanoes and earthquakes that occur among them. But volcanic and seismic activity are not confined only to those particular zones. Some volcanoes occur inside a plate's perimeter, and geologists think this may be due to hot spots. Hot spots are concentrations of heat in Earth's mantle.

As a plate moves over a stationary hot spot, molten rock (magma) may periodically rise in plumes to Earth's surface, and sometimes the plumes can create a volcano. As the plate continues to move over thousands and millions of years, one volcanic island moves off the hot spot and a new volcanic island forms directly over the hot spot (Figure 4). Where this process continues over time, a chain of volcanic islands forms. Sometimes the volcanoes that form do not rise above sea level.

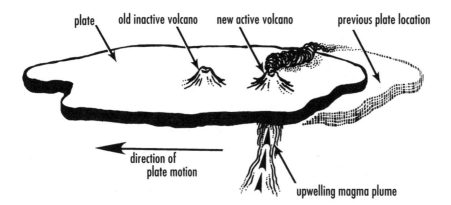

plate   old inactive volcano   new active volcano   previous plate location

direction of plate motion

upwelling magma plume

Figure 4

Magma rises through Earth's crust because it is less dense than the rock surrounding it. Generally, substances tend to rise through other substances when they are less dense. The density of a particular substance is directly related to its temperature. Water in its liquid state provides an excellent example of the relationship between density and temperature. Since hot water is less dense than cold water, it tends to rise through cold water. In this Activity a dropping bottle filled with hot water represents a hot spot in Earth's mantle; a piece of styrofoam represents a tectonic plate. The hot water rising in the dropping bottle forms a plume that rises to collide with the bottom of the styrofoam "plate," creating a "volcano." As the plate is repositioned, a new area is moved directly over the plume and a new volcano forms. The old volcano becomes inactive. As the plate continues to move, a chain of inactive volcanoes with one active volcano at the end is formed. Many geologists think that the pattern produced by the Hawaiian Islands and the Emperor Seamounts provides substantial evidence for the theory of plate tectonics.

This Activity gives students an opportunity to investigate how a hot spot can form a volcano, and how studying chains of volcanoes can provide data to determine plate motion.

Important Points for Students to Understand

◊ Not all volcanoes occur at plate boundaries. Volcanoes within plate perimeters probably result from hot spots in Earth's mantle.

◊ Volcanoes within plate perimeters usually form in chains as a plate moves over a stationary hot spot.

◊ Molten rock (magma) rises mainly because it is less dense than the surrounding rock, just as hot water rises through cold water.

◊ As with many other features related to plate tectonics, the process of volcanoes forming as a result of hot spots occurs very slowly over time.

Time Management

This Activity can be completed in approximatley 30 minutes.

Preparation

Before beginning the Activity, carefully cut the "tectonic plate" out of styrofoam using a knife. (Styrofoam used in packaging meat or vegetables may be obtained from a local supermarket, and cleaned and reused in following classes.) Each plate should fit comfortably within the width of the shoebox. It should be long enough to allow room for several hot spots, but short enough to allow for movement from one end of the box to the other. Collect the remaining materials and divide them among the trays set out for each group.

Suggestions for Further Study

Recently declassified satellite data are currently providing scientists with a wealth of new information about undersea geological formations. The data have already revealed some formations that do not fit standard theories about undersea geologic processes. As a result the hot spot model for volcanic island chain creation, among other undersea natural processes, is being debated throughout the global scientific community.

The debate about hot spots provides an excellent example of how technological advances affect accepted scientific ideas, and of how new information becomes assimilated into or alters existing theories. Have students investigate the newly released data, and encourage them to develop an understanding of how scientists use data to substantiate a scientific theory. This exercise may be used to underline student appreciation of science as an evolutionary process.

Suggestions for Interdisciplinary Study

The Hawaiian Island chain is one of Earth's most beautiful locations. What is beauty? Have students write a description of the most beautiful thing they have ever seen. Are beautiful things valuable because they are beautiful? How do people care for those things that are valuable to us, such as memories, art, or possessions that have personal meaning. How do we care for Earth?

Answers to Questions for Students

1. The hot water rose out of the dropping bottle toward the bottom of the styrofoam "tectonic plate."

2. The cold water flowed into the dropping bottle and formed a layer on the bottom of the bottle.

3. Hot water and magma are both less dense than their surrounding materials. Less dense materials rise through more dense materials.

4. A volcano.

5. A chain of volcanoes.

6. The arrow drawn by the students on Figure 3 should move from the top left toward the bottom right hand corner.

7. The X drawn by the students should be placed above and to the left of the uppermost circle (active volcano).

8. The Emperor Seamounts and Hawaiian Islands form a continuous chain of volcanoes that track across the Pacific Ocean. The volcanoes are generally oriented from northwest to southeast, with a distinct bend in the middle.

9. Recent Pacific Plate movement has been to the northwest.

10. If both the Hawaiian Islands and the Emperor Seamounts were formed by a single, stationary hot spot, then the Pacific Plate has not always traveled in the same direction. The Pacific Plate is currently traveling in a northwesterly direction. When the Emperor Seamounts were formed, the Pacific Plate was traveling in a northerly direction.

11. Answers will vary. However, students should be encouraged to identify a variety of strengths and weaknesses of this model of volcano formation. For instance, students might respond that this is a good model because the heat source is located in one place and the plates move across that source, forming a series of volcanoes. Students might also comment that the "plate" in this model is rigid like the actual lithosphere, and that the "asthenosphere" in the model has the ability to flow. It is important to be aware of the misconceptions that this model may convey, such as that the actual asthenosphere is made up of molten or liquid material.

from *The Santa Barbara Earthquake*

Way out in California
Upon a hill so tall,
Was the town of Santa Barbara
That they thought would never fall.

But on one fatal morning,
The sun rose in the sky,
The people all were praying,
"Oh, Lord, please hear our cry."

When daylight found the people
With sad and aching heart,
They were searching for their families
That the earthquake tore apart.

Vester Whitworth

Shake It Up

Background

Nearly all locations on Earth experience occasional earthquakes, although most of them are not large enough to cause significant damage. But people living near plate boundaries often experience earthquakes, and sometimes those earthquakes can cause large amounts of damage.

Damage from earthquakes could be significantly reduced if people just avoided living in areas where a lot of earthquakes occur. But that is hardly ever possible. By using certain construction designs and materials the damage caused by earthquakes could also be significantly reduced. But such designs and materials can be expensive, and city governments sometimes have to weigh the costs of using them against the possibility that an earthquake will actually occur in their area.

In this Activity you will use sugar cubes to investigate and compare the effects of an earthquake on different construction designs. You will also learn about how people make decisions when they want to construct buildings in areas where there are frequent earthquakes.

Objective

To compare various construction designs' ability to withstand the effects of an earthquake.

Materials

You will work in groups of four or more for this Activity. Each group will need
◊ two books, same size
◊ one shoe box lid or tray
◊ 20 sugar cubes
◊ a pencil or crayon
◊ a ruler
◊ Student Worksheet

Procedure

1. Use the ruler to locate the center of the shoe box lid or tray. Mark this center point with the pencil or the crayon (Figure 1). This center point will be the epicenter for the earthquakes you will create.

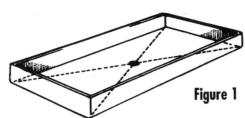

Figure 1

2. Stand the two books on their ends and rest the shoe box lid or tray on top. Space the books so the ends of the lid or tray are resting on each book (Figure 2).

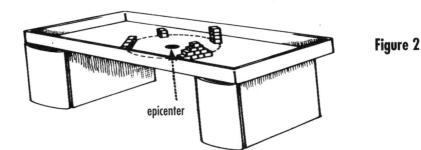

Figure 2

epicenter

3. Place the sugar cubes at various spots near the center of the lid or tray, but not on top of the end over either book. Use the following configurations

   ◊ four cubes side-by-side
   ◊ two cubes, one on top of the other
   ◊ four cubes in a stack, like a tower
   ◊ ten cubes stacked like a pyramid

The center of each sugar cube structure should be approximately five centimeters from the earthquake epicenter. An exact measurement is not required so long as the structures are all equidistant from the epicenter.

4. Lightly tap the underside of the lid or tray directly beneath the epicenter. Continue tapping lightly until one of the sugar cube structures topples. Make a note of which structure topples first under the heading Trial A on the Worksheet (page 131).

5. Continue tapping the lid or tray progressively harder until all the sugar cube structures have toppled. Note the order in which they topple on the Worksheet.

6. Repeat the trial using the same designs. Place the sugar cube structures in the same positions on the lid or tray. Record the results under Trial B on the Worksheet.

7. In Trial C, choose one of the above four designs but this time vary the distance between the structure and the epicenter. Place one structure directly over the epicenter and the others at varying distances. Record the distances from each structure to the epicenter on the Worksheet.

8. Lightly tap the underside of the lid or tray directly beneath the epicenter. Continue tapping gently until one of the structures topples. Make a note of which one topples first under Trial C on the Worksheet.

9. This time, rather than continuing the earthquake until all the structures topple, note the damage done to the other structures immediately after the first structure toppled. Did they slide? Are the cubes now misaligned? Record your findings in the Worksheet.

10. Repeat Trial C to verify your findings. Record the results under Trial D on the Worksheet.

11. Create additional earthquakes using your own experimental sugar cube structure designs. As you experiment, remember to try to alter only one variable at a time. For example, design your own structure and place a few of them at varying distances from the epicenter to see how they withstand an earthquake at the various locations. Design your own data collection sheet to record the data from your experiment so that it may be easily understood by others. Repeat the experiment as often as time allows. Record subsequent trials on your own data sheet.

12. When all trials have been completed clean your area and return the materials to their proper places.

13. Your teacher may ask each group to present its findings from their experimental trials.

Questions/Conclusions

1. Which structure toppled first during Trial A?

2. Why do you think it was first to topple?

3. Which structure toppled last during Trial A?

4. Why do you think it was last to topple?

5. Were the results for Trial B the same as the results for Trial A? If so, why? If not, how and why did they differ? Were all of the test conditions exactly the same? What changes in the conditions could have changed the results?

6. Using the data you collected from Trials A, B, C, and D, as well as the data collected from your own experimental trials, what general predictions can you make about how the structure of a building affects its stability during an earthquake? Do you think your prediction would apply to a real building? What differences between the sugar cube structures and real buildings might affect the accuracy of your prediction?

7. Building construction designs and materials that minimize the potential for earthquake damage tend to be very expensive. Using them can raise taxes and increase costs for goods and services in communities where earthquakes occur. What factors should people in high-risk earthquake areas consider when deciding how much protection against earthquake damage to include in their building construction designs?

8. Earthquakes cannot be prevented or controlled, but sometimes there are warning signs, such as bulging or tilting of the ground, increases in the number of tremors, changes in well-water levels, and sudden changes in the vibrations in the bedrock of Earth's crust. The safest place to be during an earthquake is out of doors, away from structures that could topple, crumble, or fall. Protective padding can be worn around the head and shoulders as protection against falling objects and flying debris. The safety of individuals in a community may depend upon their ability to react quickly and take the necessary precautions at the earliest sign of an earthquake. Can you think of circumstances that might hinder a person's efforts to react properly to signs of an earthquake?

9. Sugar cubes are not buildings. What are some of the weaknesses of using sugar cubes to model the affect earthquakes have on buildings? What are some of the strengths?

Trial A: Standard Designs

Structure Type	Order in which it Toppled

Trial B: Standard Designs

Structure Type	Order in which it Toppled

Trial C: Distance from Epicenter

Distance	Result/Damage

Trial D: Distance from Epicenter

Distance	Result/Damage

Shake it Up

Materials

◊ two books, same size
◊ one shoe box lid or tray
◊ 20 sugar cubes
◊ a pencil or crayon
◊ a ruler
◊ Student Worksheet (page 131)

What Is Happening?

Virtually all locations on Earth experience earthquakes at some time or other. People living near plate boundaries live in active seismic areas and they may experience earthquakes frequently. Earthquakes in such areas also tend to be more intense. Generally, the damage in a region located at or near the epicenter of an earthquake will be more severe than in other areas. However, the number, size, and types of structures located there and the degree to which the population is prepared to react to an earthquake are all important factors.

While earthquakes cannot be prevented or controlled, sometimes there are warning signs, such as bulging or tilting of the ground, increases in the number of tremors, changes in well-water levels, and sudden changes in the speed of seismic waves (waves of energy resulting from the vibrations that emanate from the bedrock in Earth's crust). The time of day or night that an earthquake occurs can be a crucial factor affecting human preparedness. A warning network may be largely ineffective in the middle of the night, for example. A building in which many people work during the day poses danger to a significant number of people during an earthquake that occurs during the day. Highway bridge buckling and damage pose particular threats during times of heavy traffic.

Damage caused by earthquakes can be reduced by not locating buildings in high-risk areas. However, economic and population factors may make building in such areas unavoidable. Certain construction designs and materials can be used to minimize the potential for structural damage. These techniques, however, tend to be very expensive, and for public buildings and highways are usually paid for through increased taxation and rises in the costs of goods and services. In deciding where and how to build, the protective value of earthquake-resistant construction must be weighed against economic costs.

In this Activity students use sugar cube structures to investigate and compare the affects of an "earthquake" on different building designs. This Activity is designed to encourage and facilitate discussion on the circumstances affecting earthquake preparedness, including weighing the costs of certain construction techniques against the likelihood of an earthquake's occurrence.

Encourage students to be aware of the various components of the scientific method they are employing in this Activity. These include: experimentation and observation, repetition, hypothesis formulation, variability, and data collection and reporting.

Important Points for Students to Understand

◊ Earthquakes occur all over the world, but they occur most frequently and with greater intensity in active seismic regions, especially along plate boundary areas.

◊ Earthquakes cannot be prevented or controlled, but some construction designs and materials will lessen the risk of property damage.

◊ For a given region, the cost of building structures that can better withstand earthquakes must be weighed against the likelihood of an earthquake's occurrence.

◊ Many circumstances may affect a person's ability to react to an earthquake in a manner that ensures their safety.

◊ Earthquake preparedness is important, especially in areas where seismic activity frequently occurs.

Time Management

This Activity can be completed in one classroom period. The first four trials offer the opportunity to discuss consistency in methods of experimentation, variability, and possible reasons for repeatability of results. Subsequent trials allow students to experiment with their own building designs, with data collection, and with hypothesis formulation and testing.

Preparation

Prior to beginning this Activity, collect the necessary materials and divide them among the groups. Before distributing the materials to each group, stress the fact that earthquakes occur everywhere around the world, but that people living at or near plate boundaries are in active seismic regions. In such areas, earthquakes occur more frequently than in other areas, and they may be especially intense. The damage that results from a given earthquake depends on many variables, including the response readiness of the population, the number and types of structures, and the materials out of which those structures are built.

Inform students that some buildings, because of their construction, sustain less damage during an earthquake than others. Such structures can be, however, more expensive to build. Explain to students that they are about to perform an experiment that will allow them to compare the types and amounts of damage that different structural designs are likely to sustain during an earthquake. At the conclusion of the experiments, have students report their findings to the class.

Suggestions for Further Study

Have students research the damage caused by some major earthquakes (such as those listed on page 21) and compare the damage done to different types of structures, or by similar structures at different distances from the earthquake's epicenter. Have students investigate the ways in which earthquakes are predicted, recognizing that earthquake prediction is usually inaccurate.

Have students research emergency preparedness and response plans in your community. By contacting local and regional emergency-response officials, as well as organizations dedicated to disaster response and victim relief such as the American Red Cross and the Federal Emergency Management Agency (FEMA), students will gain valuable insight to the relationship between civic responsibility, humanitarian aid, and emergency preparedness. The class might want to perform a mock earthquake-response drill. Have students become familiar with the warning signs of earthquakes.

Suggestions for Interdisciplinary Study

Some geometric shapes are inherently more stable than others. Architects and engineers have incorporated these shapes in designing structures that can withstand the shock waves created by earthquakes. Have students investigate what geometric shapes exhibit stability, how they are used by architects and engineers, and what mathematical principles underlie their stability.

Compare the different effects of two distinct earthquakes of similar intensity on their respective communities. Choose earthquakes that occurred during the same historic era among societies with similar construction and warning technologies (Los Angeles and Kyoto), or choose earthquakes that occurred during distinct historic eras among societies with quite different construction and warning technologies (Pompeii and Mexico City). What architectural and socio-economic factors might account for the differences?

Answers to Questions for Students

1. Usually the four-cube, stacked structure will be first to topple.

2. It is the least stable geometric design.

3. In most cases the pyramid will be last to topple.

4. It has the most stable geometric design.

5. Answers may vary. If results from Trials A and B were the same, all test conditions were the same. If results were not the same, something in the test conditions differed between the two trials. Answers will vary as to what test conditions were changed. Possible changes that could affect the results are: structures were placed in different locations; tapping was done at a different location; tapping was done with a different intensity; structures were "built" in different ways.

6. Generally, the taller or narrower a building is the less stable it will be during an earthquake. Answers will vary as to predictions for real buildings. Differences that may affect the accuracy of predictions might be: the degree of "connected-ness" of the building sections as opposed to the degree of "separatedness" of the sugar cubes; the sinking of building foundations below ground level as opposed to the setting of the sugar cubes directly on top of the shoe box lid or tray; earthquake resistant construction methods that may be used in real buildings. Flexible materials are able to move during an earthquake without breaking. Rigid or brittle construction materials usually break or crumble. Thus, wooden buildings and buildings that are properly reinforced with steel may survive an earthquake, while masonry structures (brick, concrete blocks) are often destroyed.

7. Answers will vary. Possible factors are cost, whether or not the funds are available to construct a more resistant structure; the purpose of the structure—a hospital or school may need to be especially earthquake resistant; how many people will occupy the structure at one time; the history and location, frequency and intensity of earthquake activity and the resulting damage to structures in the area.

8. Answers will vary. There are many actual and potential circumstances that could hinder proper reaction to an earthquake. Some of the most significant include:

    √ Time of day or night an earthquake occurs. A warning network might not be effective in the middle of the night, when most people are asleep. A building such as a school or office, where many people work during the day, may pose danger to more people during an earthquake that occurs during the day than one that occurs during the night. Highway bridges pose a special threat during a time of day with heavy traffic loads.

√ Crowded conditions, as in a stadium or sports complex, may prohibit people from reaching exits. Crowds under any circumstances pose dangers if panic occurs.

√ Lack of warning signs, such as early tremors, preceding a major earthquake.

√ Confinement in an enclosed area, such as a subway, bus, automobile, or room, when an earthquake occurs.

9. Answers will vary, but encourage students to list a variety of strengths and weaknesses for the model, including how buildings are actually constructed.

The finest workers in stone are not copper or steel tools,
but the gentle touches of air and
water working at their leisure
with a liberal allowance of time.

Henry David Thoreau

Rock Around the Clock

Background

Earth is over four *billion* years old. By comparison, the oldest people in the world live to be about 100. How many hundreds does it take to make a billion? Most of Earth's changes occur very slowly, and people do not readily perceive them. But the fact is Earth has evolved and changed dramatically over the past four billion years, and it will continue to evolve and change.

A single rock provides an example of how slowly geological changes occur on Earth. If you picked up a rock and kept it the rest of your life, you would probably notice little or no change. Yet rocks can and do change, it often just takes quite a long time. Many factors cause rocks to change, but what is important to understand is that change *does* occur. This Activity will give you an opportunity to investigate how rocks can change over time.

Deep within Earth, rocks can encounter temperatures high enough to make them melt. After the melted (molten) rock has cooled, it solidifies and a new rock is formed. This new rock is called **igneous rock**. **Weathering** causes rocks to break down into smaller pieces. **Erosion** causes rock fragments to be transported by wind and water to other places. As rock fragments become steadily deposited in a particular place they form layers of sediment. As the layers build up, their combined weight causes the lowest layers to compact. The tiny spaces between rock fragments fill with natural cementing agents that act like a kind of glue, or the mineral grains in the rock grow and interlock. A new type of rock is formed called **sedimentary rock**.

In some locations on Earth, tectonic plates are moving toward one another and even colliding. In such areas, rocks can be affected by the enormous pressures and temperatures associated with plate movement.

These extreme conditions can transform both igneous and sedimentary rock into a new type of rock. This new rock is called **metamorphic rock**.

These are the three types of rocks, and after any one type is formed it can be weathered, eroded, and deposited over and over again. All of these processes taken together are called the **rock cycle**.

Objective

To investigate the processes by which rocks are formed and broken down, and to see how rocks can change over time.

Materials

Each student will need
◊ safety goggles
◊ a lab apron
◊ a pocket pencil sharpener
◊ scrap paper
Each lab group will need
◊ eight wax crayons
◊ tongs
◊ two pieces of lumber 2.5x12.5x20 cm (exact sizes are not necessary)
◊ hot plate
◊ aluminum foil
◊ four envelopes
◊ newspaper
◊ vise, optional

Vocabulary

Igneous rock: Rock formed after molten rock solidifies.
Sedimentary rock: Rock formed when layers of sediment compact.
Metamorphic rock: Rock formed by the extreme pressures and temperatures caused by tectonic plates moving toward one another and colliding.
Weathering: The process by which rocks break down into smaller fragments.
Erosion: The process by which rock fragments are broken down, transported, and redeposited.
Rock cycle: All the processes of rocks forming, breaking down, reforming, and transportation taken together.

Procedure

Part 1

1. Cover all table tops with newspaper. Cover the hot plate burner with aluminum foil.

2. Trade four of your crayons with another group, so that each group has two crayons of each of the four colors.

3. Shave the crayons with the pocket pencil sharpener on a piece of the scrap paper. Keep all the shavings of each color in its own pile. Examine the shavings and record your observations. If the shavings must be stored overnight, place each pile in its own envelope. What part of the rock cycle is being simulated by shaving the crayons ? Hint: The crayons represent rocks.

Part 2

4. Fold a 30-cm-square piece of aluminum foil in half to form a rectangle. Place one color of the crayon "rock" fragments in the middle of the aluminum foil. Spread the shavings into a square layer approximately one centimeter thick.

5. Carefully spread another color of "rock" shavings on top of the first layer, forming a second layer. Do this with each remaining color so there is a four-layer stack of crayon rock fragments in the middle of the foil rectangle (Figure 1). What part of the rock cycle does this step represent?

6. Carefully fold each side of the aluminum foil over the stack of rock fragments, allowing for a one-centimeter gap between the edge of the shavings and where the foil folds.

7. Place the foil package between the two pieces of lumber, and place this "sandwich" on the floor. Apply moderate pressure by pressing it together with your hands. What part of the rock cycle does this step represent?

8. Remove the foil package from between the two boards, and carefully open it. Record your observations. Gently lift the sandwiched "rock" material out of the package with both hands, placing your fingers underneath and your thumbs close together on top. Now break the sandwich into two pieces. Dump the loose rock fragments onto a piece of the scrap paper and save them. Record your observations. Place the two parts of the rock fragments back into the foil and refold the package.

Figure 1

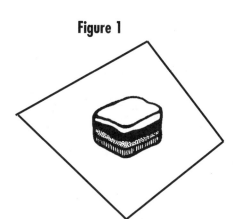

Part 3

9. Place the foil package back between the two boards and apply as much pressure as you can. A vise works best, but you can apply enough pressure by simply standing on it. What part of the rock cycle does this step represent?

10. Open the foil package and examine the newly formed "rock." Break the "rock" into several pieces and record your observations.

Part 4

11. Fold a 30-cm-square piece of aluminum foil in half and fashion it into a melting bowl large enough to contain the rock fragments.

12. Place the contents of the first foil package into the bowl.

13. *Before proceeding, make sure you have on your safety goggles and lab apron. Handle the hot plate and melted crayon rock fragments with extreme care!* Make sure the hot plate is completely covered with aluminum foil, then set it to medium temperature. Place the foil melting bowl on top of the hot plate and melt the rock fragments (Figure 2). Be careful to melt them slowly enough to keep the fragments from spattering, and stop the melting process before the fragments fuse completely together.

Figure 2

14. Turn off the hot plate, *but remember that it will still be hot for a while longer before it cools!* Use the tongs to carefully remove the bowl from the hot plate and set it aside to cool for about ten minutes. *Be extremely careful not to spill the molten rock fragments on anyone or anything!* What part of the rock cycle does this step represent?

15. Once the rock has thoroughly cooled, remove it from the bowl. Break it open, examine it, and record your observations.

Questions/Conclusions

1. Name and describe the parts of the process by which rocks break down. At what stages of this Activity did each occur?

2. Explain how a sedimentary rock might become a meta-morphic rock.

3. Briefly explain how this Activity relates to the rock cycle. What are the strengths and weaknesses of this Activity as a model for how the rock cycle works?

4. How likely is it for you to see what was simulated in this Activity within your own lifetime? What is probably the only type of rock you might have an opportunity to see formed in your own lifetime? Explain your answers.

Rock Around the Clock

What Is Happening?

Earth is constantly changing, yet people are unable to perceive most of Earth's changes because those changes occur over such long periods of time. Many students may think that a rock is always a rock, so to speak, and have no idea that a rock they pick up after school today hasn't always appeared the way they see it. Rocks can and do change over time, and this Activity is designed to illustrate those changes. Conceptually, geologic change over a vast expanse of time can be difficult for students to comprehend. This Activity focuses on the main components of the rock cycle, and enables students to investigate that cycle by using a physical model to simulate tectonic forces.

There are three primary rock types: igneous, sedimentary, and metamorphic. Igneous rock is formed when magma cools at or beneath Earth's surface. Much sedimentary rock forms as a result of the compaction of layer upon layer of sediment deposits. Both igneous and sedimentary rock can be changed by extremes in temperature and/or pressure into metamorphic rock.

All three rock types are subject to weathering and erosion. Weathering occurs both chemically and mechanically. Chemical weathering involves the change of one mineral into another. Mechanical weathering involves the breakup of minerals and rocks through the physical actions of wind and water, among other natural forces. Erosion means weathering coupled with physical transportation of rock and mineral fragments, such as by rainwater runoff. Weathering and erosion play important roles in the formation of sedimentary rock because they are the agents by which sedimentary layers are deposited on top of one another. Weathering and erosion are the principal agents by which all three rock types are broken down at Earth's surface. Below Earth's surface rocks are broken down primarily through extremes of heat and pressure.

Such temperatures and pressures are most often associated with converging plate boundaries, where two or more tectonic plates are moving toward one another or colliding. The combination of high temperature and great pressure in such areas can cause any type of rock to be changed, even metamorphic rock. Tectonic forces can later lift metamorphic rock to Earth's surface, where it is subject to weathering and erosion, or those

Materials

Each student will need
◊ safety goggles
◊ a lab apron
◊ a pocket pencil sharpener
◊ scrap paper

Each lab group will need
◊ eight wax crayons
◊ tongs
◊ two pieces of lumber 2.5x12.5x20 cm (exact sizes are not necessary)
◊ hot plate
◊ aluminum foil
◊ four envelopes
◊ newspaper
◊ vise, optional

forces can plunge it deeper into Earth where it may become magma.

The rock cycle can become fairly complex. Yet the basic processes are straightforward: some forces combine to form rock materials while other forces combine to break them down. All the processes taken together are called the rock cycle, and the rock cycle occurs over geologic, as opposed to human, time. This Activity enables students to investigate the rock cycle through a physical model that simulates tectonic forces.

Figure 1
The rock cycle

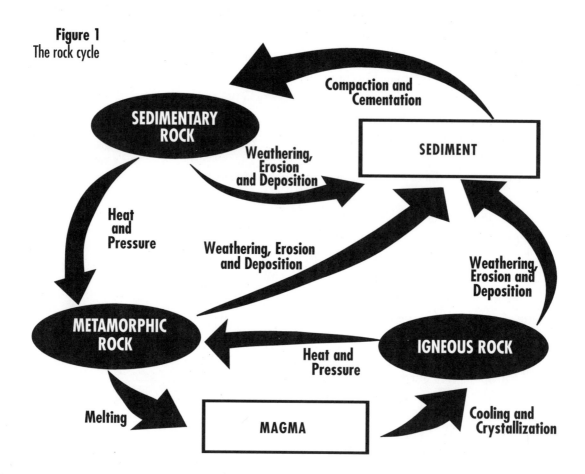

Important Points for Students to Understand

◊ The rock cycle occurs slowly over geologic time. Most changes occur so slowly that people are incapable of noticing them with their five senses.

◊ Rocks on Earth's surface are broken down through chemical and mechanical weathering.

◊ Most sedimentary rocks are formed by compaction of rock fragments beneath Earth's surface.

◊ Extreme pressure, usually found at converging plate boundaries, causes changes in rock that result in metamorphic rock.

◊ Igneous rock is formed when magma cools and solidifies.

Time Management

This Activity will take approximately three class periods. On Day 1, introduce the Activity and complete Part 1. On Day 2, complete Part 2, Part 3, and Steps 1-4 of Part 4. On Day 3, complete Step 5 of Part 4, and the Conclusion. Allow for discussion and cleanup time.

Preparation

No special preparations are required for this Activity.

Suggestions for Further Study

Chemical and mechanical weathering not only affect rocks and minerals, but are the agents by which all Earth's components are broken down. The fundamental difference between chemical and mechanical weathering is that the former produces new (different) minerals, while the latter results in smaller particles of the original minerals. Have students think about the effect that chemical and mechanical weathering have on various different substances, like wood, animal carcasses, and rocks, in order to reinforce their comprehension of the relative expanse of geologic time. Emphasize that while chemical and mechanical weathering affect different substances at different rates, they are nevertheless the primary ecological agents by which all Earth's organic and inorganic matter are broken down.

Suggestions for Interdisciplinary Study

Intertwined with the study of geology are many issues relevant to the quality of human life and the well-being of Earth. Addressing such issues can prove a difficult task. By encouraging students to write expressively about such topics, both teachers and students will have an opportunity to confront and discuss some of the most important issues of our time. Strip mining, for example, is a subject involving issues such as land use and management, sustainable resources and conservation, pollution, habitat destruction, and recycling. Some of the social and economic issues related to this subject include sources of mined materials, who mines them, who buys those materials, who profits from their sale, and who is harmed if there is a ban on mining. Students might select and research one or more of these topics and write an essay or put together a presentation. Students may also be encouraged to take a position on a controversial issue, and explain why they have chosen their particular stance.

When teaching middle-level students abstract concepts such as plate tectonics and rock formation, it may prove beneficial to provide a way for students to relate an abstract concept to a concrete object. The Earth Materials List below matches common objects with the raw materials from which they are produced. As a focus activity, bring some of these objects to class and play 20 Questions to see if your students can identify the mystery objects.

EARTH MATERIALS

aluminum can	aluminum ore (i.e., mineral bauxite)
automobile	iron ore (i.e., magnetite) aluminum ore, quartz
record album/CD	petroleum
gold jewelry	gold ore
silver jewelry	silver ore
tin can	iron ore, tin ore
glass	mineral: quartz
roads	rock, sand, petroleum
clothes (some)	petroleum
plastics	petroleum
pennies	minerals: cuprite, bornite (copper)
table salt	mineral: halite
building materials	
concrete	sand and rocks (i.e., limestone and shale)
bricks	clay and sand
steel	iron ore, titanium ore, manganese ore

Answers to Questions for Students

1. Weathering. Chemical weathering can occur when a chemical change takes place in which bonds between substances in the rock are weakened or broken. Mechanical weathering occurs when a physical force acts upon a rock causing it to break, such as rain, wind, and cold. Weathering was simulated in this Activity by shaving the crayons.

2. As sedimentary rock on one plate collides with that on another plate, enough pressure can be exerted to cause it to change into metamorphic rock.

3. In Part 1, the crayons being shaved into small pieces represents the weathering process. The placement of the shavings in layers in Part 2 represents the deposition of rock fragments into sedimentary layers. The pressure placed on the crayon shaving models the compaction of sediment caused by the weight of the sedimentary layers above, which creates the sedimentary rock. The increased pressure in Part 3 relates to how metamorphic rocks are formed. The heating of the crayons in Part 4 represents the changes that occur as a result of rock material melting during the formation of igneous rock. Student answers regarding the model's strengths and weaknesses will vary. Encourage students to list a variety of each for the model, including points such as the similarities and differences between rocks and crayons, and how long it takes for rocks to be altered through the rock cycle compared to how long it takes for crayons to melt and reform, among others.

4. It is not likely that any of the changes within the rock cycle simulated in this Activity will be observed by students within their lifetimes because of the extreme lengths of time required for such processes to occur. Probably the only rock people might have the opportunity to actually see being created is an igneous rock formed as a result of volcanic activity, such as that at Mt. St. Helens or in Hawaii.

from *The River Duddon, Sonnet XV*

FROM this deep chasm, where
    quivering sunbeams play
Upon its loftiest crags, mine eyes
    behold
A gloomy NICHE, capacious, blank,
    and cold;
A concave free from shrubs and
    mosses grey;
In semblance fresh, as if, with dire
    affray,
Some Statue, placed amid these
    regions old
For tutelary service, thence had
    rolled,
Startling the flight of timid
    Yesterday!

William Wordsworth

Study Your Sandwich, & Eat It Too !

Background

About three-fourths of Earth's land areas are made up of sedimentary rock, rock that forms in layers. Geologists obtain information about the history of a particular area by studying the area's **rock formations**. A rock formation is what geologists call a distinctive unit of rock that formed in a particular area. Geologists study rock formations so they can learn about how an area formed over geologic time.

Rock formations are easy to study when they are exposed, like along road cuts or cliffs. But most rock formations are not exposed, and lie deep within Earth. Geologists have had to develop special techniques to study them. **Core sampling** is one technique geologists use to study hidden rock formations. Core sampling means drilling into a rock formation and pulling out a specimen of all the rocks that have formed there. By pulling up a number of different specimens from a certain area, geologists can construct a chronology (time line) of geologic events for that area.

In this Activity you will use your sandwich as a model for a rock formation made of sedimentary layers. You will learn about the special technique of core sampling, and about how geologists use the information they get from core sampling to construct geologic time lines for specific areas.

Procedure

1. Pick up a tray of materials from your teacher.

2. The ingredients used to create your sandwich represent layers of rock. Name each ingredient to represent a layer of rock. For example, white bread could represent sandstone, and chunky peanut butter could represent a conglomerate.

   ✳ Record the name in Table 1 on the following page, and create a graphic symbol for each layer. You will use these symbols later in the Activity to complete the diagrams.

Objective

To investigate the core sampling techniques geologists use to collect information about rock formations, relative ages, and faulting.

Materials

You will work in groups for this Activity. Each group will need
- ◊ one slice white bread
- ◊ one slice whole wheat bread
- ◊ one slice dark rye bread
- ◊ two tablespoons jelly
- ◊ two tablespoons chunky peanut butter mixed with raisins
- ◊ two paper plates
- ◊ plastic knife
- ◊ measuring spoon
- ◊ clear plastic straws

Vocabulary

Rock formation: A distinctive rock unit that formed in a particular geographic area.
Core sampling: A technique, used by geologists to get information about rock formations, involving drilling into Earth and pulling out specimens of rock from a formation's many layers.

TABLE 1		
Ingredient	Name	Symbol

* 3. Place the white bread on a paper plate. Next, spread chunky peanut butter onto the white bread. Add the whole wheat bread, the jelly, and the rye bread. Your sandwich represents a rock formation with five layers of rock (Figure 1). How do the layers in your sandwich differ? How do you think rock layers differ?

Figure 1

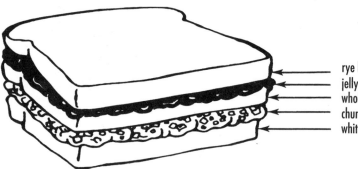

rye bread (layer 5)
jelly (layer 4)
whole wheat bread (layer 3)
chunky peanut butter with raisins (layer 2)
white bread (layer 1)

4. Important points for geologists to establish when studying a particular rock formation are the relative ages of the various rock layers. In the rock formation represented by your sandwich, which layer represents the oldest rock layer? Where is it located? Explain your answer. (Hint: what does it mean to be the oldest?)

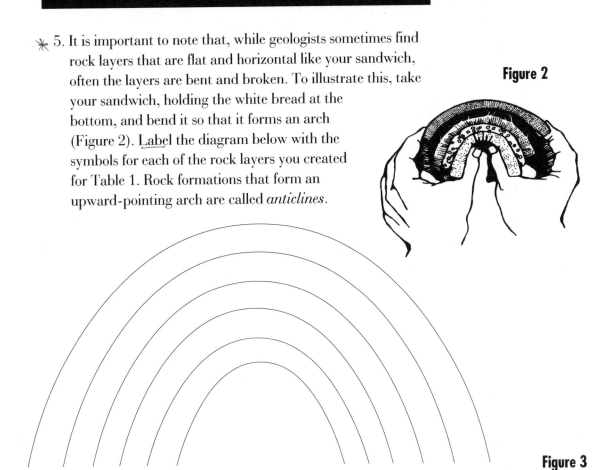

Figure 2

✳ 5. It is important to note that, while geologists sometimes find rock layers that are flat and horizontal like your sandwich, often the layers are bent and broken. To illustrate this, take your sandwich, holding the white bread at the bottom, and bend it so that it forms an arch (Figure 2). <u>Label</u> the diagram below with the symbols for each of the rock layers you created for Table 1. Rock formations that form an upward-pointing arch are called *anticlines*.

Figure 3

✳6. Now bend your sandwich downward to form a trough, again holding the white bread at the bottom (Figure 3). <u>Label</u> the diagram below with your rock layer symbols from Table 1. Rock formations that form a downward-pointing arch are called *synclines*.

Figure 4

7. Geologists are concerned with the relative ages of the various rock layers within the overall rock formation, and with the sequence of events which has affected the formation. Which is the oldest layer in your rock formation? Which is the youngest? Was the formation deformed—bent upwards or downwards—before or after the original layers formed?

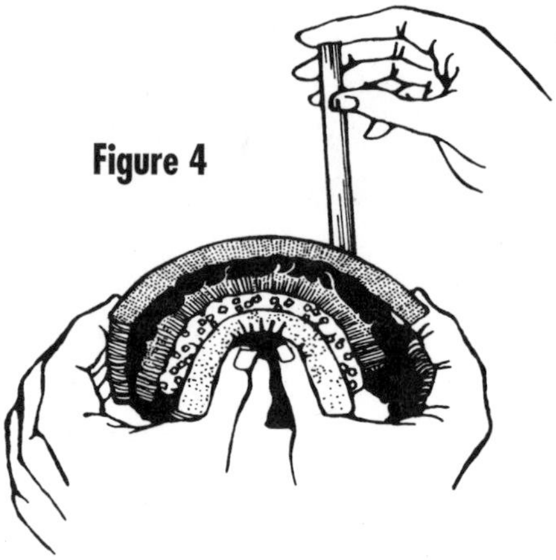

8. Keeping the white bread at the bottom, one member of the group should carefully bend the sandwich again to form an arch. Other members of the group should then use clear plastic straws to take three core samples at the points indicated on the diagram at left. Push the straw into the sandwich from top to bottom, making sure to keep the straw in a vertical position, regardless of whether the rock layers in your sandwich are horizontal or not (Figure 4). Make sure that you keep the core oriented correctly so that you remember which is the top of the core sample. After you have finished taking the three core samples, sketch the results of your samples in the empty cylinders provided below, labeling the diagram using the symbols you created from Table 1.

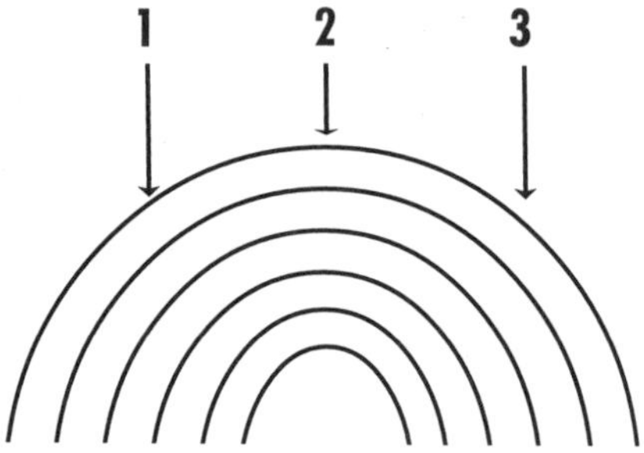

1   2   3

Example

1   2   3

✳ 9. Repeat Step 8, but this time bend the sandwich downward to form a trough. Sketch the results of your core samples, using the appropriate symbols from Table 1, in the empty cylinders below.

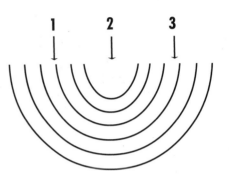

1   **2**   **3**

10. Sometimes rock formations can be completely broken, rather than just bent. To simulate this, cut your sandwich in half. Holding one of the halves in each of your hands, move one half up or down relative to the other. Do the layers match up anymore? Draw a side view of what you ✳ observe in the space below.

11. What you have just created with your sandwich halves is a simulation of a vertical fault. Movement of rock layers along a vertical fault can cause earthquakes. Another type of fault is called a lateral fault. A lateral fault can be simulated with your sandwich model by sliding the two halves past one another at the same level. Movement of rock layers along lateral faults can cause earthquakes as well. In the space provided below, draw a side view and a top view of your lateral fault.

12. Again, observe your sandwich. Which is the oldest rock layer? Did the faulting occur before or after the original layers were formed?

Questions/Conclusions

1. Your sandwich serves as a model of the way a rock formation might appear. How effective is this model in portraying core samples of Earth's interior? What other kinds of models could be used?

2. Describe the order of events which produced the samples seen in the cores you took in Step 8. Which layer was formed first, second, and so on, and when did significant deformative events occur?

3. Describe the order of events which produced the results observed in Step 10.

4. How are the chronologies described in the above two questions similar? How are they different?

5. For the set of core samples shown in Figure 6, create a cross-section diagram in the space provided below of what the rock formation from which they were extracted might look like. Describe a possible chronology of events leading to this particular rock formation.

Figure 6

6. How accurate a picture could you make if you had only one of the core samples shown in Figure 6? What does your answer tell you about the importance of having a lot of data or making a lot of observations before reaching a conclusion?

7. Why might it be important for geologists to locate faults?

8. For what reasons might it be important to know what is under Earth's surface?

Study Your Sandwich & Eat It Too !

What Is Happening?

This Activity is designed to give students an opportunity to investigate core sampling, one of the techniques geologists use to study rock formations. Students will also learn how geologists use the data they get from core sampling to construct a chronology of geologic events in a particular area.

Rock formations are not always readily available to geologists for direct visual study. In such situations, core sampling is a technique geologists use to gather information about what lies beneath Earth's surface. As with all sampling techniques, the number of samples and the locations where they are taken are important variables in determining how accurate a picture can be reproduced using the samples.

In establishing the order of events in a rock formation's geologic chronology, it is important to consider the relative ages of the layers of rock present. The oldest layers are found at the bottom and the youngest layers are found at the top, unless the sequence has been overturned or disturbed in some way. Also, horizontal layers that are bent or broken indicate a disruptive event that occurred after the layers were formed. A sense of the chronology of geologic events is an important concept for students to learn and appreciate.

This Activity presents a highly simplified illustration of the many natural forces involved in creating rock formations. Emphasize to students that this is just an introduction. There are many complexities to core sampling, since rock layers can be overturned, weathering and erosion can remove entire sections of a formation, and not all formations occur in horizontal layers.

Important Points for Students to Understand

◊ Sedimentary rocks generally form in horizontal layers, with the oldest layer on the bottom. The layers may be disrupted or even overturned by later events.

◊ Disruptive events may bend or break sedimentary layers after they are formed. The forces that cause such events may also cause earthquakes.

Materials

You will work in groups for this Activity. Each group will need

◊ one slice white bread
◊ one slice whole wheat bread
◊ one slice dark rye bread
◊ two tablespoons jelly
◊ two tablespoons chunky peanut butter mixed with raisins
◊ two paper plates
◊ plastic knife
◊ measuring spoon
◊ clear plastic straws

◊ Core sampling is one technique geologists use to learn about rock formations that are not exposed for direct study. As in all areas of scientific study, the number and quality of samples has a significant effect on the accuracy of the conclusions drawn from the data.

◊ This Activity provides a model for demonstrating the relative ages of rock layers, and the chronology of disruptive geologic events.

Time Management

This Activity may be completed in one class period.

Preparation

No special preparations are required for this Activity. The bread you select should be firm, day-old bread, or something similar. Fresh, soft breads do not hold up well enough to produce clear core samples. You might want to practice with the bread you're planning on using, and even have students practice taking core samples several times before proceeding with the graphic representations.

Suggestions for Further Study

Anticlines are layers of rock that, roughly, form an upside-down U shape, while synclines are those that form a rightside-up U shape. A common misconception is that anticlines are the rock configurations that create hills and mountains, while synclines are those that create valleys. However, this is not true in all cases. The type of rock configurations found beneath Earth's surface does not always directly coincide with the topography of the above landscape. Have students research the rock formations in your local area and investigate the relation of those formations to the landforms under which they lie. Pay particular attention to the chronology of geologic events that has led to present day landforms.

Actual core sampling requires considerable effort to drive sampling machinery through Earth's layers. Have students investigate how far down geologists have been able to extract samples. Use this investigation to reinforce their understanding of geologic distances and the actual depth from Earth's surface to Earth's core layers.

Suggestions for Interdisciplinary Study

Determining a sequence of events is important not only in geology but also in many areas of physical science and of human history. Learning to analyze evidence and data in order to produce a plausible sequence of events is a skill that takes practice in order to perform well. Figure 7 shows two sets of animal tracks. Have students look at the tracks, and see if they can determine what happened to the two animals using the evidence provided. Have them write a narrative explaining their determination, and construct a time line based on the chronology of their narrative.

Figure 7

Answers to Questions for Students

Procedure #3. Student answers might include that the layers in their sandwiches differ in color, texture, and consistency, among other differences. Encourage students also to note that the layers differ in their relationships to one another, such as their distances, heirarchy, and what separates them. Rock layers differ for many of the same reasons, but respective ages is among their primary differences.

Procedure #4. The oldest layer is on the bottom and the youngest layer is on top. To be the oldest means the bottom layers formed before any others.

Procedure #7. The oldest layer is on the bottom and the youngest layer is on the top. The disruption of layers occurred after the layers were formed.

1. Answers will vary. Students may not clearly understand how geologists take core samples, but encourage thought along the lines of the sampling issues addressed in Activity 7. Students may also recognize the limitations imposed by current technology related to core sampling, such as depth. Encourage students to understand that the ages of a rock formation's layers are different from those of their sandwiches. Other kinds of models might include the use of clay, actual sediment, and computer models, among others.

2. Answers will vary depending upon the names of the various rock layers. The bottom layer is the oldest and the top layer the youngest. The folding to produce the anticline occurred after the horizontal rock layers were formed.

3. Answers will vary depending upon the names of the various rock layers. The bottom layer is the oldest and the top layer the youngest. The break in the formation to produce the vertical fault occurred after the horizontal rock layers were formed.

4. The chronologies are similar in that both begin with horizontal rock layers being formed, followed by a disruptive event that altered the rock formation. The chronologies are different in the type of disruptive event that occured.

5. Answers will vary depending upon the names and symbols of the various rock layers. The bottom layer is the oldest and the top layer the youngest. The disruption of the rock formation layers to produce the anticline occurred after the horizontal layers were formed.

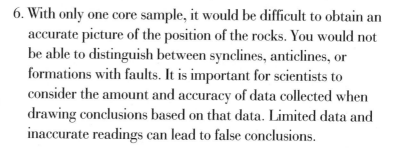

6. With only one core sample, it would be difficult to obtain an accurate picture of the position of the rocks. You would not be able to distinguish between synclines, anticlines, or formations with faults. It is important for scientists to consider the amount and accuracy of data collected when drawing conclusions based on that data. Limited data and inaccurate readings can lead to false conclusions.

7. Earthquakes are created by movement of rock formations along faults. Also, valuable minerals are sometimes found along faults.

8. Answers will vary. Possibilities include locating valuable resources, determining where earthquakes might occur, and understanding how Earth is constructed.

from *A poetical geognosy*

When nature was young, and Earth in her prime,
All the rocks were invited with Neptune to dine.
On his green bed of state he was gracefully seated,
And each as they enter'd was civilly greeted.
But in choosing their seats, some confusion arose,
Much jostling and scrambling, and treading on toes.

First Granite sat down, and then beckon'd his queen,
But Gneiss stepp'd in rudely, and elbow'd between,
Pushing Mica-slate further; when she with a frown
Cried, 'You crusty, distorted, and hump-back'd old clown!'
But this was all sham,— for to tell you the truth,
They had been the most intimate friends from her youth,
But let scandal cease. See the whole tribe of Slates
All eager and ready to rush to their plates;
Oh heav'ns! how the family pour in by dozens,
Of brothers, and sisters, and nephews, and cousins!
The elder-born Limestones ran in between these,—
They were very well known to be fond of a squeeze.

John Scafe

Rocks Tell a Story

Background

Identifying rocks can be extremely difficult, even for geologists. Proper rock identification depends on the quality of the specimen and on the clarity of its significant characteristics. Through this information, rocks provide clues about the environmental conditions under which they were formed. Geologists use that information to reconstruct an area's geologic history. This Activity will give you an opportunity to study various rock samples, and to suggest probable reasons why certain rocks have certain characteristics.

Throughout this Activity, it is important to remember that a rock is a mixture of different **minerals**. The specific minerals and their relative concentrations—how much of each mineral appears in the rock—are responsible for the rock's characteristics, including its color, density, and pattern of cleavage.

Keep in mind the three major rock types—igneous, sedimentary, and metamorphic. In general, specimens of the same kind of rock (granite, an igneous rock, or sandstone, a sedimentary rock, for example) will exhibit similarities in characteristics, such as texture, **crystal** size, color pattern, and density. However, variations in characteristics exist between individual specimens, between different kinds of rocks within the three major rock types, and between the major rock types themselves. A rock's characteristics tell geologists about its composition, and about the conditions under which it formed.

Procedure

1. From the information provided in the rock-sample set, complete the first two columns of Table 1 on the following page.

2. For each rock, record your observations of the specimen in Table 1 on the following page. Pay particular attention to the specimen's color, density, texture, and other characteristics that may provide clues as to the rock's geologic history.

Objective
To observe the characteristics of rock specimens.

Materials
Rock-sample set containing:
◊ 1. gabbro and 2. basalt
◊ 3. slate and 4. shale
◊ 5. granite and 6. gneiss
◊ 7. sandstone and 8. conglomerate
◊ 9. limestone and 10. marble

Vocabulary
Mineral: Elements or compounds that occur naturally in Earth's crust; building blocks of rocks.
Crystal: A mineral with a specific geometric shape (i.e. a cube or prism).

TABLE 1			
No.	Name	Type	Characteristics
1			
2			
3			
4			
5			
6			
7			
8			
9			
10			

3. Now compare specimens 1 and 2. These two rocks are related in some way. How are they similar? How are they different? Compare rocks 3 and 4, rocks 5 and 6, rocks 7 and 8, and rocks 9 and 10 in the same way. Record your observations in Table 2 on the opposite page.

4. For each pair of rocks, suggest reasons for the similarities and differences you recorded in the tables. Write down your thoughts, and be prepared to share your ideas as part of a class discussion.

5. Each pair of rocks are related in some way. What story can you create to explain their relationship? Be prepared to share your story with the class in some form, such as written or spoken.

TABLE 2

Rock Pair	Differences	Similarities
slate & shale 1 and 2		
3 and 4		
5 and 6		
7 and 8		
9 and 10		

Rocks Tell a Story

Materials

Rock-sample set containing:
◊ 1. gabbro and 2. basalt
◊ 3. slate and 4. shale
◊ 5. granite and 6. gneiss
◊ 7. sandstone and 8. conglomerate
◊ 9. limestone and 10. marble

What Is Happening?

This Activity is designed to serve as a wrap-up in which students are asked to examine rock specimens to determine differences and similarities between selected pairs. This Activity relies on a student's ability to make accurate observations, and to make use of previous experiences with rocks and minerals from preceding Activities and other sources. The most important question for students to think about is: What do the characteristics of a particular rock tell us about the conditions under which the rock was formed and about the composition of the rock?

In this Activity, it is important to de-emphasize the classification and naming of the rocks; this is not an exercise about naming rocks only. Students should focus on the characteristics of an individual specimen rather than try only to determine its name. Each group should be provided with a rock-sample set that clearly displays the rocks' names, along with their classification within the three major rock types: igneous, sedimentary, and metamorphic.

The factors that go into determining a rock's characteristics are complex. This Activity is designed to serve only as a means for opening a discussion of how geologists use information about rocks to reconstruct geologic history.

Important Points for Students to Understand

◊ A rock's type (igneous, sedimentary, or metamorphic) provides information about the conditions under which the rock was formed.
◊ Mineral grain size and shape provide additional details about the events that led to the formation of a specific rock.
◊ The color of a rock and its mineral composition provide information about the chemical composition of a rock.
◊ Two rocks of the same composition can have quite different appearances. The differences indicate that the two rocks did not form in exactly the same way, meaning they have different geologic histories.

Time Management

This Activity can be performed in one class period.

Preparation

No special preparations are required for this Activity. Rock-sample sets may be purchased from Earth science or geological supply companies (See Appendix B, page 200).

Procedure

Once students have completed and recorded their observations, ask them to suggest reasons for the similarities and differences they observed between each rock pair. Initiate a class discussion by encouraging responses. Allow students to suggest different hypotheses to explain the similarities and differences they observe. As discussion proceeds use the following information to assist students in evaluating their ideas.

   #1 & #2. Gabbro and basalt are both igneous rocks formed directly from molten material. They have essentially the same composition. Some samples of basalt have cavities formed by gas bubbles in molten rock. Mineral grains or crystals can be identified in gabbro, indicating a slow underground cooling process. The basalt contains only very tiny mineral grains, indicating a fast cooling process. Both rocks typically are found in oceanic crust.

   #3 & #4. Shale is a sedimentary rock and is one of the parents of the metamorphic rock called slate. Both specimens may have similar colors. The density and hardness of the slate indicate that the shale from which it formed was subjected to intense pressure. The pressure caused tiny mineral grains to line up in parallel formations, and the slate breaking along those grains resulted in its flat smooth surfaces.

   #5 & #6. Granite is an igneous rock and is one of the parents of the metamorphic rock called gneiss. Mineral grains are evident within the granite, giving rise to its rough texture and also indicating a slow cooling process. The mineral grains within the gneiss are visible but may appear deformed. Within the gneiss some distorted layering may be visible, providing further evidence that metamorphic forces shaped the rock. Note that small samples make it difficult to see the layering. Gneiss and granite may have the same composition and therefore the same color, or the composition and color may differ.

#7 & #8. Sandstone and conglomerate are both sedimentary rocks, formed from loose rock, mineral, or fossil fragments that were cemented together. The most obvious difference between these two specimens is the fragment size. These relative sizes give some indication as to the environment in which the sediments were originally deposited. The conglomerate might indicate a stream or river channel, while the fine-grained sandstone might indicate a coastal region where sands are deposited.

#9 & #10. Limestone is a sedimentary rock and is the parent rock of the metamorphic rock called marble. Marble's higher density indicates it has been subjected to intense pressures. Crystal growth generally is evident within marble. The limestone usually is a relatively soft rock. Marine fossils are often found within limestone layers, indicating their formation in warm marine environments. Metamorphism usually destroys the fossils when the limestone is converted into marble.

Suggestions for Further Study

Have students investigate other related rock types, focusing on the characteristics of the rocks which indicate something about their geologic history. Have students collect local rocks, especially those that show signs of change (for example, rock samples showing progressively more weathering or rocks with a range of mineral grain size).

Suggestions for Interdisciplinary Study

Storytelling has been an important component of human culture for centuries, especially before the age of widespread literacy. Stories were passed orally from generation to generation to perpetuate traditions, cultural identity, and social history. How has the art of storytelling changed over time? Have students interview older Americans, such as grandparents or neighbors, to ask their views on this subject. Ask students to bring to class a story passed along to them by an older relative or friend that they might share with the class.

Introduction

The following readings elaborate on the concepts presented in these Activities. Although they were written especially for this volume with the teacher in mind, students should be encouraged to read them for both interest and additional study.

Plate Tectonics

This is a very brief overview of what geologists today refer to as plate tectonics. The older term, "continental drift," is perhaps better known and was used by earlier theorists who developed the idea that the continents had not always been fixed in their current positions. The modern theory of plate tectonics states that the outer part of the Earth consists of relatively thin, rigid pieces called plates, and that these plates continually, if very slowly, move. It might be helpful to think of Earth as a hard boiled egg whose shell has been cracked into pieces that move over the egg white's surface. Today it is generally accepted among geologists that there are six large plates and a number of smaller ones moving over Earth's surface (Figure 1). Each plate is about 100 kilometers thick and has an area of thousands of square kilometers. At present the plates are moving at rates ranging from 1-20 cm/yr.

Figure 1

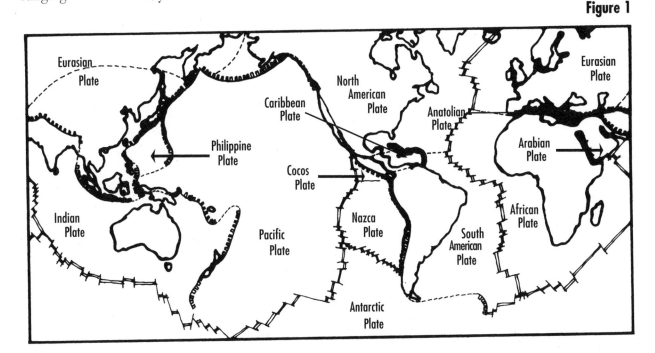

A Unifying Theory

The theory of plate tectonics helps explain many geological phenomena, especially the existence and distribution of earthquakes, volcanoes, and mountain ranges. Plate tectonics provides a unifying theory that explains large-scale geologic change over expanses of time, helps us understand why Earth's geological formations currently appear the way they do, and helps scientists predict changes in Earth's form that may occur in the future.

Modern world maps hold the same continental movement clues early theorists noticed from the maps that charted European naval exploration after the 15th century. Francis Bacon wrote in the early 1600s about the "good fit" between the coastlines of "old" and "new" worlds and, over 200 years later, similarities among certain plant fossils collected by Antonio Snider in Europe and America supported Bacon's observation. Snider also recognized the apparent fit between continental coastlines and, in trying to determine the geographic domain of fossil plants, he pieced the continents together like a jigsaw puzzle. Snider theorized that the continents had once been a single, huge landmass. Other hard evidence of continental movement came in 1885 when Eduard Suess found the same rock formations among several southern-hemisphere continents. The formations were identical in sequence and type, so they almost certainly formed at the same time and place. Since that could not have happened among continents always separated by oceans, Suess theorized that they must have once been joined. He, like Snider, theorized a single historical landmass.

Continental drift was first proposed as a distinct scientific theory by F.B. Taylor and Alfred Wegener in the early 1900s. Their hypotheses, developed independently of one another, were based on evidence similar to that discussed above and included the suggestion of a single historical landmass. While the theory of continental drift attracted much attention, the insufficiency of hard evidence and the absence of an explanation about how the continents moved at all hindered its acceptance among the international scientific community. Although the theory of continental drift was not generally accepted in the early 1900s, continuing curiosity and improving technology provided additional evidence in its support. Two examples are:

Figure 2

Fossil records. Remains of an extinct fern seed were found in South Africa, Australia, and India. Because the seeds were large, scientists thought it unlikely they could have been dispersed by wind or water. This implies that these three continents had once been closer together, if not actually joined.

Fossil remains of an extinct reptile, mesosaurus, were also found distributed among distinct areas of eastern South America and western Africa (Figure 2).

<u>Paleomagnetism</u> (ancient magnetism). Some types of rocks are weakly magnetic. When they were being formed certain iron-bearing minerals within them became aligned parallel to Earth's magnetic field, like a compass's needle; they "pointed" to magnetic north. After these rocks crystallized, the iron bearing minerals were "frozen" in place, and today they show the direction of magnetic north at the time they were formed. The frozen, internal "compass needles" of rocks formed at the same time among other continents show the direction of magnetic north at the same time in history for each continent. How does such data support the theory of continental movement? Figure 3 presents three hypothetical continents—A, B, and C; the arrows represent the direction of magnetic north for each continent. Note that the arrows do not point toward a common magnetic north. In Figure 3, it appears that magnetic north was in three different places at the same time, or that there were three entirely different magnetic norths. Yet, if the continents had moved since the rocks were formed, it would take only a single magnetic north pole to explain the current differences. In Figure 4, the hypothetical continents have been repositioned so the arrows point toward a common magnetic north.

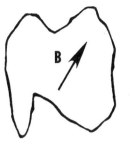

Figure 3

New Discoveries

So far we have looked at several types of evidence that appear to support the theory that continents have moved. Overwhelming evidence for continental movement—now called plate tectonics—began to accumulate in the 1950s. With more sophisticated technology, scientists began systematic exploration of the world's ocean floors. While most of these scientists were not deliberately seeking data to either support or disprove continental movement, their findings became crucial to the development and general acceptance of the theory of plate tectonics.

Figure 4

They discovered a huge system of undersea mountains

formed by molten rock. Branches of this system occur on all the world's ocean floors, and rise as much as three kilometers. Running along the crest of each of these mountain ranges is a valley-like feature called a *rift valley* (Figure 5). A rift valley is a crack or series of cracks in Earth's crust; in the Atlantic and Indian Oceans, they are associated with mountain chains that follow a path along the ocean floor roughly midway between the continents. For this reason, many of these ranges are called mid-ocean ridges. The Mid-Atlantic Ridge is probably the best known. In a few places, these ridge-and-rift systems cut across continents: the Red Sea lies in a huge rift valley.

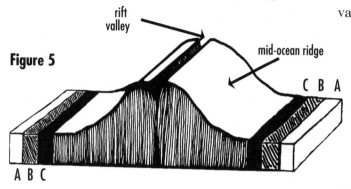

Figure 5

rift valley

mid-ocean ridge

C B A

A B C

In the early 1960s, further exploration of seafloor geological features revealed even more evidence to support the plate tectonic theory. As one example of this new evidence, the age of rocks from the seafloor was found to increase moving away from the ridges and toward the continents. In Figure 5, samples of rock obtained by drilling at Location C were found to be younger than samples taken from Location A and Location B. Those from B were younger than those from A. This is because the mid-ocean ridges and the seafloors are igneous, formed by molten rock from Earth's interior that has periodically risen to the surface, cooled and hardened. At an area of weakness in Earth's crust—in this case a weakness in the rift dividing the mid-ocean ridge into two parts—magma is able to rise to the surface. This may be more easily understood if we assume three discrete episodes of movement of magma through Earth's crust and examine the results (Figure 6). From

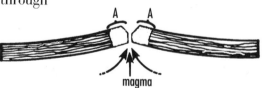

A   A

magma

Figure 6

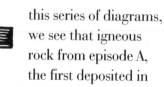

A B   B A

magma

A B C   C B A

spreading

this series of diagrams, we see that igneous rock from episode A, the first deposited in our example, moves away from the crest of the ridge when magma intrudes during episodes B and C. This phenomenon is

called *seafloor spreading*. The seafloor is moving away from the rift as a result of additions of new magma to the oceanic crust. Specific evidence for this process is provided by the relative ages of the rocks that form the seafloor, as well as by the magnetic patterns preserved in those rocks. As new rock is added to the oceanic crust, it records the characteristics of Earth's magnetic field at the time. As oceanic crust moves in both directions away from the ridge, the magnetic pattern also shifts. The spreading sea floor has acted something like a tape recorder, preserving a magnetic record that tells us the history of continental growth and movement.

Figure 7

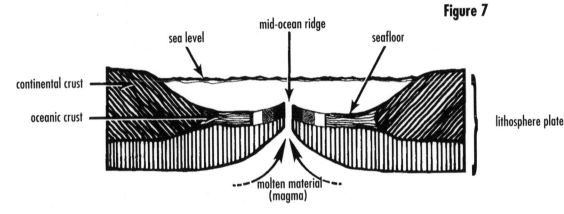

The relationship of the seafloor, continents, and Earth's interior is shown above in a more detailed sketch of Earth's crust (Figure 7). The continental and oceanic portions of Earth's crust form part of the lithosphere, the relatively rigid outer shell of Earth. When new oceanic crust forms at mid-ocean ridges both the older oceanic crust and the plate move away from the ridge. Continents riding atop the plates move along with them.

Convection

After years of accumulating evidence that lithosphere plates have moved in the past, more recent measurements show that the plates are still moving. Scientists are not in complete agreement as to what causes plate motion, but one suggestion is that convection currents within Earth's interior provide the driving mechanism. Many scientists think convection occurs in the asthenosphere due to heat generated from Earth's interior. A convection current occurs when a liquid or gas is heated, becomes less dense, and rises. When cooling makes the liquid or gas dense again, it eventually sinks. This cyclical process forms a convection cell, with material within the cell continually moving in response to heat energy being released into the material.

Convection currents also can occur in hot solids, although their motion is very slow. Beneath the lithosphere is a portion of Earth's interior called the asthenosphere. The rock in the asthenosphere is rigid but, because it is extremely hot and under intense pressure, it does not behave like the rigid rocks found on Earth's surface. The rock in the asthenosphere has fluid characteristics and flows in various directions. The asthenosphere's fluid motion moves the lithosphere plates and the continents that ride atop them. Rifts, or cracks, appear periodically in the plates and become filled with new magma that rises from Earth's interior.

Even though ocean floors are continually spreading and new oceanic crust is continually being removed, Earth is not growing larger. While exploring the Pacific Ocean floor in the 1950s, scientists often found deep, narrow trenches quite near a continental coast, such as the one that parallels the western coast of South America. Trenches occur where one plate (the more dense one) sinks beneath another plate into Earth's interior (Figure 8), where the material that forms the sinking plate melts and becomes part of the asthenosphere. Figure 8 shows the relationship between convection cells, plate motion, mid-ocean ridges, and trenches.

Figure 8

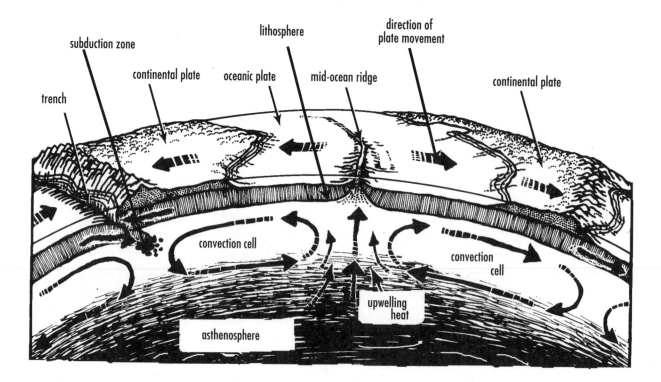

Slab Pull

geologists that convection currents within the asthenosphere act as the primary driving force of plate tectonics has recently been disputed by some scientists offering an alternative theory. Proponents of this alternate theory suggest that two forces play a significant role in causing plate movement: ridge push and slab pull. Both forces are related to gravitational and thermal effects. Scientists theorize that hot material within the mantle rises beneath oceanic ridges carrying not only mass to Earth's surface but heat as well. This heat causes the oceanic lithosphere to expand, thus creating an elevated ridge system. The result is that the youngest edge of the oceanic lithosphere rests on an inclined plane above the easily deformable asthenosphere below. Gravity causes the edge of the plate to slide down this inclined plane and push on the rest of the plate in the direction of plate motion, hence "ridge push." As the plate edge moves away from the ridge it also moves away from the heat source. Consequently, it subsides and contracts as it cools, causing the entire plate to become colder, thicker, denser, and heavier. Eventually it might become gravitationally unstable and plunge, or be subducted, into the mantle. As it is subducted, the low melting point fraction of the lithosphere will be boiled off (rich in water), and with increasing depth there may occur phase changes in some minerals into denser forms. The cold, dense, subducting plate will literally fall into the mantle, pulling on the remainder of the plate, hence "slab pull."

Boundary Types

Now that we have seen how plates are created and recycled, let us consider in more detail their edges or boundaries. There are three boundary types.

Figure 9
Diverging plate boundary

Divergent boundary. Occurs where two plates are moving away from each other, as along the mid-ocean ridges (Figure 9). New crust is formed at such boundaries by the upwelling of molten material from the Earth's interior. Earthquakes and volcanoes are most frequent in such areas.

Convergent boundary. Occurs where plates move *toward* each other (Figure 10). One plate moves beneath the other into Earth's interior and a submarine trench is formed. If a continent rides on the edge of one of the colliding plates, the plate without the continent is more dense and will move beneath the continent-bearing plate, becoming reassimilated into the asthenosphere. Earthquakes and volcanoes may occur in such areas. The collision of two plates, each with a continent riding close to its edge, may result in the formation of a mountain range. The Himalayas were formed when the plate carrying India collided with the plate carrying China.

Figure 10

Converging plate boundary

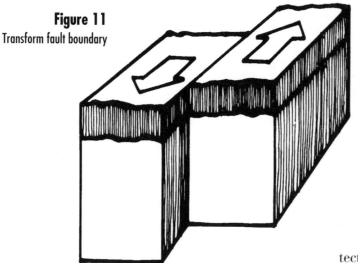

Transform fault boundary. Occurs where there is no significant movement of two plates toward or away from each other. There is, however, *lateral* movement between the two plates (Figure 11). The San Andreas fault in California is a transform fault boundary. The friction generated by its lateral movement caused major earthquakes in San Francisco in 1989 and in Los Angeles in 1994.

Figure 11

Transform fault boundary

The theory of plate tectonics provides a unifying explanation for many geological phenomena that in the past were not well understood. These include volcanic and earthquake activity, the formation of large mountain chains, and the many features of the ocean floors. Investigation continues and will, scientists hope, result in an even better understanding of how and why Earth "works."

Volcanoes

In our review of plate tectonics we noted that volcanic activity is associated with a boundary where two plates converge. While not all volcanic activity is related to plate boundaries (some volcanoes occur in the center of plates, and some plate boundaries exhibit no volcanic activity at all), most of these frequently spectacular, sometimes destructive volcanic eruptions are located at plate boundaries.

Volcanoes are formed when molten rock (magma) emerges through an opening, or vent, from Earth's interior. Such an opening may be above or below water. Lava is magma that has been forced out of Earth's interior onto the surface. The accumulation of volcanic lava and rock debris around the opening forms a cone- or dome-shaped structure. Figure 1 shows a simplified representation of the relationship between these basic geologic features.

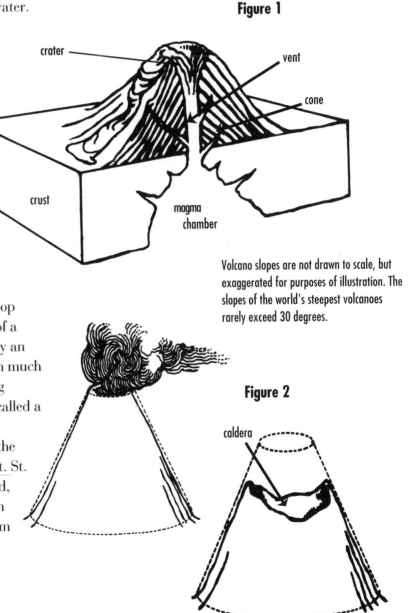

Figure 1

crater

vent

cone

crust

magma chamber

Volcano slopes are not drawn to scale, but exaggerated for purposes of illustration. The slopes of the world's steepest volcanoes rarely exceed 30 degrees.

A *crater* is a steep-sided, funnel-shaped depression at the top of a volcano. Sometimes the top of a volcano collapses or is removed by an eruption, resulting in a depression much larger than the previously existing crater. This geological feature is called a *caldera* (Figure 2).

Volcanoes have two forms: the conical strato volcano, such as Mt. St. Helens or Vesuvius, and the broad, gently sloping shield volcano such as Mauna Loa. The volcano's form is determined by what emerges during an eruption.

Figure 2

caldera

Volcanic Eruptions

The nature of a particular eruption, as well as the shape of the resulting cone, depends upon the proportion of gaseous, liquid, and solid material produced, plus the chemical composition of the magma. Following is a brief summary of volcanic products.

Gases. Volcanic gases are primarily water vapor (steam) and carbon dioxide. Other volcanic gases include hydrogen sulfide and compounds of chlorine, fluorine, and boron. The "smoke" frequently seen emerging from a volcano is actually a combination of condensing steam and volcanic dust.

Liquids. The liquid product of volcanoes is lava. Lavas vary in chemical composition and physical properties, and their chemical composition affects their viscosity, which in turn affects how fast and how far they flow. Lavas are classified according to their silica (SiO_2) content: *felsic* (65-75% silica), *intermediate* (50-65% silica), and *mafic* (<50% silica). Felsic lavas tend to be quite viscous while the mafic lavas are less viscous.

Solids. Solid volcanic products are called *pyroclastics*. These represent fragments of solidified magma within the volcano vent that are blown out when gas pressure increases. Pyroclastic fragments are usually classified by size, and range from *volcanic ash* (<2 mm) to *volcanic blocks* or *bombs* (>64 mm).

While silica content determines the viscosity of lava, water content provides the potential for steam pressure that determines explosive force. Following is an examination of silica and water content combinations to show how each affects eruption behavior and cone shape.

Low water/low silica. Usually results in a relatively quiet, non-violent eruption due to low viscosity. This is due to runny lava and the relatively low amount of water available for conversion to steam. The cone will probably be broad-based and have relatively gentle slopes, forming a shield volcano.

High water/high silica. Usually results in volcanoes with an explosive history. High viscosity inhibits the escape of steam bubbles until the rising magma is close to, or breaks through, Earth's surface. When this happens, steam bubbles expand, throwing out large amounts of volcanic material in a violent display. The steep-sided cone of a strato volcano tends to result from alternate eruptions of lava and pyroclastics.

Zones of Activity

Most volcanic activity occurs in areas where plates meet. If we plotted the locations of volcanic activity on a world map we would find that most volcanic activity occurs in three distinct zones. The first zone runs almost continuously around the edge of the Pacific Ocean, incorporating New Zealand, the Philippines, Japan, and the Aleutians, and extending along the coasts of North, Central, and South America. Because of its roughly circular shape, this zone of volcanic activity is called the "Ring of Fire," and delineates the boundary of the Pacific Plate.

Another zone of volcanic activity extends east-west through the Mediterranean Sea into Asia. The third main zone extends from Iceland southward through the middle of the Atlantic Ocean, roughly paralleling the edges of the ocean-bordering continents. This zone occurs where two plates are diverging. Magma rising between the plates has resulted in the undersea volcanic mountain range known as the Mid-Atlantic Ridge.

Figure 3 depicts three different geological situations where plate movement may result in volcanic activity. When two plates converge beneath an ocean, as in A, one plate moves beneath the other toward the asthenosphere and remelting occurs. This

Figure 3

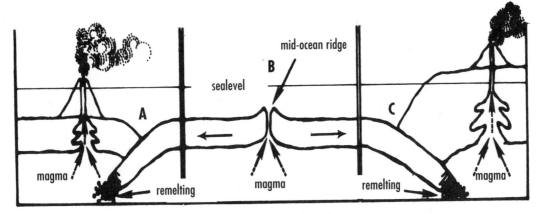

remelting may trigger volcanic activity, resulting in the formation of islands like the Aleutians in the northern Pacific Ocean. The mid-ocean ridges, as in B, are chains of volcanic mountains formed by the intrusion of magma between two diverging plates. The Azores in the Atlantic Ocean are an example of such a geological formation, volcanic mountains whose peaks protrude above the ocean's surface. When a plate moves beneath a thicker continent-bearing plate, as in C, and remelting occurs, volcanic activity may result even as far as 150 km inland. The Cascade Mountains in the Pacific Northwest are an example of volcanic mountains formed in this manner.

Hot Spots

Since the early 1900s, 90 percent of Earth's major volcanic eruptions have occurred among these three zones. An exception has been continual activity from Kilauea in Hawaii since 1924. Given what geologists now know about the relationship between plate boundaries and volcanic activity, the Hawaiian Islands, which are of volcanic origin, present an anomaly. Because they are located far from any plate boundary, geologists believe that the Hawaiian Islands were, and are still being, formed above a hot spot. Hot spots beneath a lithosphere plate result in rising plumes, or columns, of magma that may penetrate a plate and result in a volcano. Hot spots are thought to be stationary, while the plate above is thought to slide across them, as in Figure 4.

Figure 4

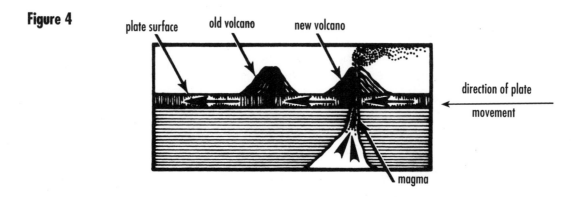

plate surface   old volcano   new volcano

direction of plate movement

magma

Figure 5

If the plate were to move, as in Figure 5, the volcano depicted in Figure 4 would have moved "off" the hot spot and a third volcano would form. Continuing movement of the plate in a particular direction would result in a chain of volcanic mountains along the route of the plate's motion, with present-day volcanic activity occurring only at the point on the plate directly above the hot spot. The reasons for the existence of hot spots are not well understood. Other hot spots exist in the Pacific, Atlantic, and Indian oceans.

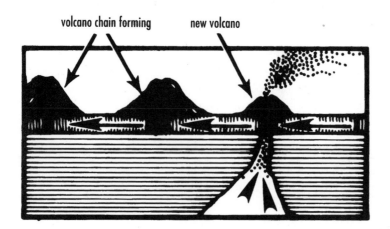

volcano chain forming   new volcano

Impact on Humans

Traditionally, volcanoes have been understood to have had a negative impact on people throughout history. One of the most devastating modern eruptions in terms of loss of life occurred on the island of Martinique in 1902. Mount Pelee, only seven kilometers from the urban center of St. Pierre, had been quiet for 200 years. In February residents noticed a sulfurous odor. In April Pelee discharged a huge cloud of ash, and eruptions accompanied by loud rumbling grew more and more frequent. By then some 2,000 of St. Pierre's 30,000 residents had sought safety elsewhere, but an equal number of people had moved within city limits from the countryside seeking refuge. During the first days of May, a series of eruptions resulted in accumulations of ash and dust approaching half a meter in thickness. A ship's captain anchored in the harbor who had seen Mt. Vesuvius erupt in Italy in 1895 set sail immediately, but few heeded his warning. On May 8th Mount Pelee erupted violently. A superheated cloud of volcanic ash and gases rushed down its slope at about 160 km/hr. St. Pierre lay directly in its path, and all but two of the 30,000 inhabitants were killed by suffocation and burning.

Deaths from volcanic eruptions result primarily from burns inflicted by hot ash and/or suffocation by ash and gases. Contrary to popular belief, lava usually moves too slowly to pose a serious threat, although it does cause extensive property damage. Mount Pelee erupted laterally, directing its force horizontal to Earth's surface. Generally, lateral eruptions cause significantly greater damage than vertical eruptions.

An example of the destructive potential of lava occurred on the Icelandic island of Heimaey in 1973. Without any prior warning, a rift opened in the ground less than two kilometers from a small town called Vestmannaeyjar, and volcanic material began to spew out in large quantities. Fortunately, the town's 5,000 residents were able to evacuate quickly. While there was no loss of life, nearly 1/3 of Vestmannaeyjar was completely buried in lava. What began as a relatively small lava flow from a single rift ended four months later as a 225-meter-high volcano cone! Named Eldfell, this volcano is an example of a rift eruption.

Another volcano was born in southwestern Mexico in 1943. Mount Paricutin started off as a fissure in a farmer's cornfield, but one week later it was a volcanic cone 168 meters high. Within a year it had grown to 335 meters. During nine years of activity Paricutin ejected an estimated 3.6 billion metric tons of ash and lava. The 4,000 residents of a nearby village moved in time to escape personal injury, but all lost their homes to lava flows.

While studying volcanic processes and hazards in order to prevent disaster is pretty straightforward—identifying past behavior, monitoring present conditions, interpreting products and processes, and empowering community leaders—shortfalls of time, funding, and logistics make it far from simple. Beginning in 1990 at the start of the International Decade for Natural Disaster Reduction, the International Association of Volcanology and Chemistry of the Earth's Interior targeted a set of high-risk volcanoes for intensive interdisciplinary investigation. By focusing work on a few volcanoes through team projects, this international endeavor is designed to reduce risks to human populations by increasing scientific knowledge about global volcanic processes.

The 1980 Mt. St. Helens explosion provides examples of other hazards that result from volcanic activity.

- Wildlife and habitat destruction. An estimated 7,000 large game animals and thousands of small animals and birds perished around Mt. St. Helens.
- Natural resources destruction. The timber blown down might have built nearly a quarter of a million homes.
- Agricultural destruction. Seven percent of some crops as far away as 150 kilometers were destroyed when volcanic dust coated leaves, inhibiting photosynthesis.
- Cleanup. Yakima, population 51,000, spent ten weeks and $2.2 million removing ash. It lies 130 kilometers away.

Throughout history people have understood volcanoes mostly in terms of their destruction of human life and works. Yet primeval volcanic episodes may have provided the initial source of Earth's air and water, and volcanoes may still generate 1/4 of all its oxygen, hydrogen, carbon, chlorine, and nitrogen. Volcanic ash is nitrogen and phosphorous rich, a major contributor to soil fertility. Volcanoes generate terrestrial landscapes on which ecosystems take root and evolve. Volcanic emissions are important sources for industrial materials like pumice, ammonia, and boric acid. Thermal energy is harnessed for industrial and domestic uses; most homes in Reykjavik, Iceland, are heated by volcanic springs, and geothermal steam produces electricity in Italy, the U.S., Mexico, Japan, and New Zealand. Remnants of volcanoes such as the Hawaiian Islands and Cascade Mountains provide opportunities for recreation and scenic enjoyment. While volcanoes are powerful natural forces that inspire awe—and sometimes fear—because of their destructive potential, they are nonetheless an integral component of Earth's ecological cycles.

Earthquakes

If you slightly bend a thin strip of metal and then release the pressure the strip will return to its original shape. The metal underwent temporary, or elastic, deformation. If the strip is bent to a greater degree, it might retain its bent shape after pressure is

Figure 1

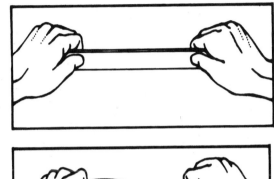

released. If so, the metal underwent plastic deformation. Should you bend the strip far enough it will break or rupture altogether. When you bend metal or wood, the energy required to bend it is stored in the object; when the object breaks, that energy is released (Figure 1).

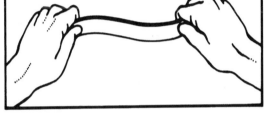

Almost all substances behave the same way, including the rocks that comprise Earth's crust. The sudden release of energy resulting from the failure of deformed rock within Earth's crust can cause earthquakes. (Earthquakes are more often caused by movement along an existing fault or break in a rock formation.) In either case, the released energy is transmitted to adjacent rock, causing it to vibrate and shake. Rock deformation is caused by the movement of lithosphere plates. Most earthquakes occur near plate boundaries where plates converge or diverge. Eighty percent of the world's recorded earthquakes occur in a zone bordering the Pacific ocean, a zone that

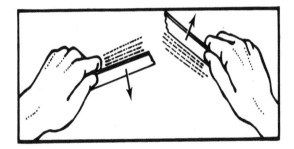

corresponds to the boundary of the Pacific plate. About 15 percent occur in a zone that stretches through the Mediterranean Ocean across India and Asia. Other geologic areas where significant earthquake activity occurs are the mid-ocean ridges, such as those in the Atlantic, Pacific, and Indian oceans. Earthquakes occur in those same areas, along plate boundaries, that have high incidences of volcanic activity.

In the United States, earthquakes often occur in California along the San Andreas Fault and along many other West Coast fault systems, and within the New Madrid Seismic area in the Midwest. Seismic activity also occurs in South Carolina's earthquake region on the East Coast and in New England. In 1989 the Loma Prieta earthquake in San Fransisco, which registered 7.1 on the Richter scale, damaged much of the downtown area, and was observed by millions on national television when it brought a World Series baseball game to an abrupt halt.

If the force applied to rock is sufficiently strong the rock will crack or break, resulting in a fracture or fault. One or both of the

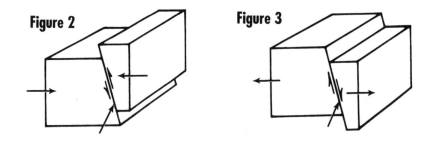

Figure 2

Figure 3

blocks of rock adjacent to the fault may move. Figures 2 and 3 demonstrate the vertical movement in Earth's crust that results from both compressional—pushing toward one another—and extensional—pulling away from one another—forces acting on a portion of Earth's crust. Lateral movement also occurs, as in Figure 4. If the rocks along an existing fault are subjected to stress, the rocks may suddenly slip, and this also results in an earthquake. The San Andreas Fault is one of the best known examples of the lateral type of fault.

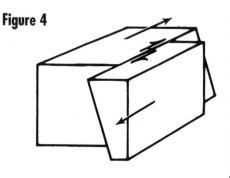

Figure 4

The subsurface point at which the rock has fractured and movement has occurred is the focus of the earthquake; the point on Earth's surface directly above the focus is the epicenter. The release of energy, which constitutes the quake, occurs at the focus and radiates in all directions as waves, called seismic or earthquake waves. These vibrating waves become less and less powerful as they move away from the focus because their energy is dissipated. The epicenter of an earthquake is the point on Earth's surface closest to the focus, and is generally where seismic waves will be strongest. Figure 5 shows the relationship between the focus, the epicenter, and seismic waves.

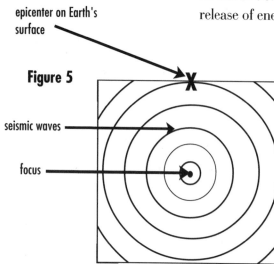

epicenter on Earth's surface

Figure 5

seismic waves

focus

Detecting Earthquakes

An estimated one million earthquakes occur throughout the world each year, although most cause insufficient disturbance to be noticed by casual observers. Earthquake waves are detected and recorded by an extremely sensitive device called a seismograph. Attached to a base is a drum holding a recording chart. The drum rotates at a predetermined speed, 2 cm/min for example. The base is firmly anchored to part of Earth's crust so that when the crust vibrates, the base also vibrates. A pendulum hangs from a scaffold that rises from the base, essentially a heavy weight to which is attached some type of marking device, such as a pen. This weight is suspended by a wire or spring above the drum so the pen comes into contact with the recording chart on the drum. Because of inertia, the suspended weight will remain stationary during an earthquake. The chart, however, will vibrate under the stationary pen causing a "zigzag" to be reproduced on the paper that indicates the arrival and force of seismic waves. Knowing the rate of the drum's rotation, geologists can determine the time interval between waves. Figure 6 shows some basic seismograph types.

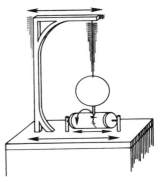

There are three kinds of seismic waves. While each type has many important characteristics that are useful in learning about the interior structure of Earth, for our purposes we will focus only on those characteristics

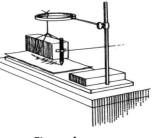

Figure 6

useful in determining that an earthquake has occurred and in locating its epicenter. The three basic wave types and their characteristics are:

- Primary Waves (P-waves). The most rapidly moving (7-8 km/sec) of the three. P-waves arrive at a given location before the other types. They travel deep within Earth's interior, and may move through both solids and liquids.
- Secondary Waves (S-waves). The second fastest, arriving at a given location after the P-waves. S-waves also travel deep within Earth's interior, but they travel through solids only.
- Long Waves (L-waves). Also called surface waves, L-waves move slowest of the three, about 3 km/sec. L-waves travel just beneath Earth's surface, and are the most destructive because they cause the surface to undulate.

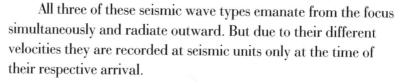

All three of these seismic wave types emanate from the focus simultaneously and radiate outward. But due to their different velocities they are recorded at seismic units only at the time of their respective arrival.

The first indication that an earthquake is occurring is provided by a series of short squiggles on the seismographic chart that represent the arrival of the P-waves. The S-waves are recorded next as a series of larger zigzags. Lastly, an even more pronounced set of zigzags indicates the arrival of the L-waves. In actual practice, two seismograph units are set up at right angles to one another to register north-south and east-west components of horizontal seismic movements. A third unit measures vertical movement. By noting the difference between the arrival times of the P-waves and S-waves, reference tables of data obtained from

Figure 7

Determing an earthquake epicenter using concentric circles plotted around three seismograph stations

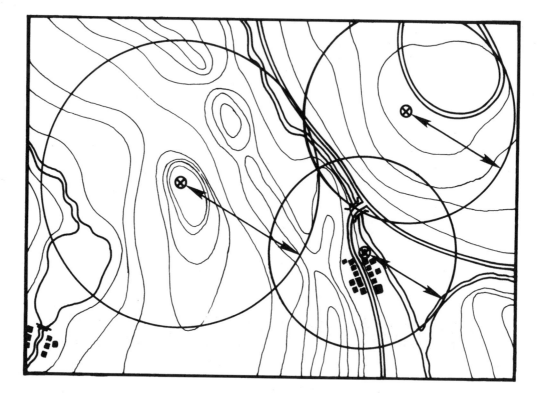

thousands of previous earthquakes are consulted to determine how far away the focus was. If three or more widely spaced seismic units are able to determine their individual distances from the epicenter, concentric circles drawn around each will locate the epicenter by determining the point where the three circles intersect. Figure 7 shows how an earthquake epicenter is determined using data from three seismograph units.

Measuring Earthquakes

Many methods for measuring earthquakes have been devised over the last century, but the two most commonly used today are the Richter and Mercalli Intensity scales. The Richter scale, devised in 1935 by the American Charles Richter, provides a system for evaluating an earthquake's magnitude by measuring the amplitude—the height of the zigzag—of seismic waves as they are recorded at seismograph units. The Mercalli Intensity scale, originally developed in 1902 by the Italian Giuseppi Mercalli, provides a system for evaluating the impact of earthquake vibrations on natural and built environments. Both of these scales provide systems for comparison, and do not yield measurements in absolute terms the way a thermometer does.

Using the Richter scale, the smallest earthquake a person might feel would be assigned a value of about 2. Because the Richter scale is logarithmic, each unit of increase (from two to three or from four to five, for example) corresponds to a tenfold increase in the earthquake's magnitude. Magnitude also can evaluate the energy released by an earthquake. Using the Richter scale, each unit increase represents a thirty-fold increase in released energy.

Using the Mercalli Intensity scale, many different variables—including magnitude, focus depth, local geological conditions, and local construction practices—are combined to assign a number between Roman numerals I and XII to rank relative levels of destruction, ground motion, and impact on humans. Because intensity is more useful than magnitude as a measure of destructive impact, professionals in the construction and property

Figure 8

RELATIVE ENERGY RELEASES		
Richter Magnitude	TNT Equivalent	Example
1.0	6 ounces	
2.5	63 pounds	
4.0	6 tons	small atomic bomb
4.5	32 tons	average tornado
5.5	500 tons	Massena, New York, quake (1944)
6.5	31,550 tons	Coalinga, California, quake (1983)
7.0	199,000 tons	Hebgen Lake, Montana, quake (1959)
8.0	6,270,000 tons	San Fransisco, California, quake (1906)
8.5	31,550,000 tons	Anchorage, Alaska, quake (1964)

sliding

Figure 9

contracting

expanding

shaking

bouncing

insurance industries rely on the Mercalli scale. However, because the Richter scale is independent of observer location, it provides a feeling for the relative size of an earthquake that may have occurred in another part of the world. This is why the Richter scale has been so readily adopted by international media. Figure 8 should provide some perspective about the relative energy released by earthquakes.

It is fair to say that the ability to accurately predict earthquakes does not yet exist. The best a geologist, seismologist, or other earth scientist can do today is predict that at some point during the next decade it is very likely that an earthquake will occur along the San Andreas fault, for example. Even with the most modern technology, we are as yet unable to predict the time, the place, or the magnitude of an earthquake.

As building construction technologies have improved over the last 50 years, there has been considerable effort among the architecture and engineering professions to devise building designs that are more resilient to earthquakes. Figure 9 shows some of the ways seismic waves affect buildings. While no construction design is ever absolutely quake proof, given the unpredictable nature of earthquake forces, many new techniques can increase human safety and reduce property damage. Generally speaking, if an earthquake affects an area built on soft sediments, such as landfills or sedimentary basins, the destruction will be greater than one affecting an area built on solid rock. Figure 10 shows some earthquake-resilient construction designs. Devices that cushion or separate a structure from its foundation have been devised to better absorb earthquake shocks. Buildings can also now be constructed with shear or interior walls that provide additional support,

Figure 10

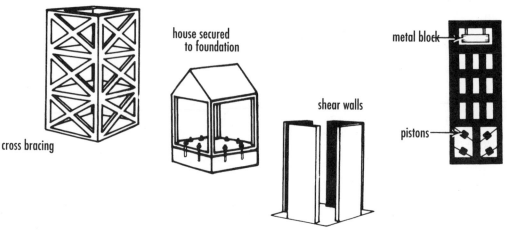

cross bracing

house secured to foundation

shear walls

metal block

pistons

and construction materials can be designed to "give" rather than resist earthquake tremors and waves.

Earthquakes of varying intensities can and do occur in many parts of the United States. It's a good idea to be aware of local emergency contingencies, and to have safety response plans ready to be put into immediate action in both the classroom and home environments. What is the status of earthquake preparedness in your community? Teaching students about why earthquakes happen provides an excellent opportunity to teach them about civic responsibility, about humanitarian aid, and about organizations that engage in disaster relief. The Red Cross and the Federal Emergency Management Agency (FEMA), for example, are excellent resources, and will provide free informational material on different aspects of emergency preparedness and response for classroom use.

Rocks and Minerals

Developing an understanding of the relationship between the most obvious aspects of geology—soils, rocks, and minerals—and the grand unifying theory of plate tectonics can be difficult. Plate tectonics describes how, and by what mechanisms, Earth's surface has been and is being transformed on a large scale. We often think of continental shifts, mountain building, volcanoes, and earthquakes as those aspects of geology explained through plate tectonics. However, rock and mineral types, locations of mineral deposits, and rock formation and transformation can also be better understood when studied in the context of plate tectonics. Plate tectonics provides a link between individual rock, mineral, and soil specimens studied in the classroom and the geologic history that has led to their development.

Minerals

A mineral is a unique combination of chemical elements, such as oxygen, silicon, aluminum, and iron. Different combinations and arrangements of elements give each mineral characteristic properties, such as color, hardness, shape, density, and cleavage. Although about 3,000 different minerals have been identified, most are extremely rare. About 95 percent of Earth's crust is comprised of only 10–15 different minerals. Some of the more common are feldspar, quartz, mica, olivine, garnet, pyroxenes and amphiboles, clay, calcite, and dolomite. This relatively small list of minerals comprises the primary rock-forming ingredients.

Rocks

Different types of rocks form in different ways. *Igneous rocks* result from the solidification or crystallization of molten rock (magma) and may form at or below Earth's surface. Two common igneous rocks are granite and basalt. They differ in their mineral composition and mineral grain size. Granite forms several kilometers below Earth's surface, where the insulating properties of the surrounding rocks cause the magma to cool slowly. This allows the growth of large mineral grains. Basalt forms at or close to Earth's surface, where rapid heat loss results in relatively small mineral grains. *Sedimentary rocks* form from rock fragments, minerals, or fossils that are compressed beneath the weight of overlying sediments. They also form by chemical precipitation of minerals dissolved by water. Sedimentary rocks form at low temperatures at or very close to Earth's surface. Common sedimentary rocks are sandstone, limestone, and shale. *Metamorphic rocks* are those in which the original mineral composition, grain size, or grain shape has changed as a result of exposure to moderately high temperatures, high pressure, or both. Most metamorphic rocks form below Earth's surface. Marble, slate, and schist are common metamorphic rocks.

The Rock Cycle

Over time, or in response to changing conditions, rocks may change from one type into another. The rock cycle depicted in Figure 1 summarizes the steps by which transformation of one rock type into another can occur. The counter-clockwise arrows indicate changes in rocks and rock materials that can take place. The words next to them refer to processes through which the changes are accomplished. Shortcuts, indicated by interior arrows, may occur in this cycle. For example, igneous rocks can be metamorphosed to form metamorphic rocks by heat and pressure or sedimentary and metamorphic rocks may undergo weathering to become sediments and/or dissolved mineral material.

Figure 1
The rock cycle

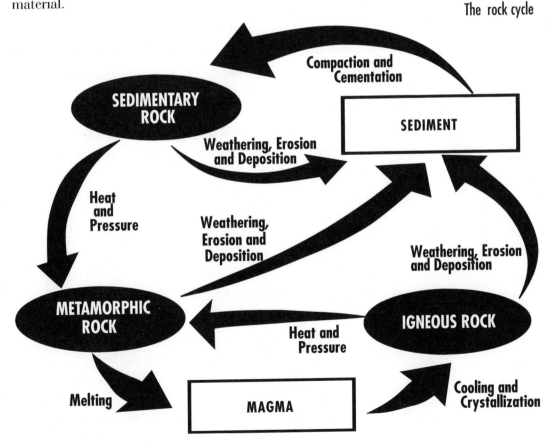

The complexity of the rock cycle suggests the difficulty in identifying and characterizing the geologic history of any single rock or mineral specimen. Strictly speaking, you must first know what minerals are in a rock before you can identify that rock. Geologists have a very difficult time identifying minerals and rocks that are weathered. It is usually easy to tell when rocks and minerals are weathered—they may crumble when handled and their surfaces may be dirty or covered with what appear to be brownish stains. In some cases, it is possible to identify the minerals in a weathered specimen, but this usually cannot be done with any degree of certainty even by someone with considerable experience. Typically, a geologist examining a weathered rock will break it with a hammer to expose a fresh surface. The geologist will then, as with any unknown specimen, examine its mineral grains using a magnifying glass if the grains are small. Finally, the geologist will make educated guesses as to the minerals present, and assign the rock a provisional name. While this initial inspection can provide some clues as to a rock or mineral specimen's composition and identity, it is generally only through the use of a specialized microscope, often aided by chemical analyses, that minerals and rocks can be accurately identified.

The identity of a rock or mineral can provide clues as to the type of environment in which the rock was formed. Generally, Earth's rocks were formed in an environment much different from the surface. For example, a typical granite that is now exposed at Earth's surface probably crystallized at a temperature of about 700°C, at high pressures, and at a depth of at least 4–10 kilometers. Such a rock, found at the surface today, became exposed as the rocks that once covered it were removed by weathering and erosion as that part of Earth's crust was uplifted in response to environmental changes. Coral-containing limestones found today on land and in relatively cold climates were initially formed in warm, shallow seas.

The identity of a rock or mineral specimen also provides clues to its geologic history and how it was affected by the motion of tectonic plates. Different rock types form at different types of plate boundaries. For example, igneous rocks are currently forming at both divergent plate boundaries (mid-ocean ridges) and at convergent boundaries. Because they are formed at plate boundaries, existing igneous rock samples can sometimes reveal the position and nature of ancient plate boundaries.

At subduction zones (one type of convergent boundary), one

plate slides beneath another and seafloor sediment in the underlying plate is carried along with oceanic crust down into the asthenosphere where it is remelted. Some of this material may eventually recrystallize into new igneous rock. Heat that is released as the molten rock cools and recrystallizes can alter adjacent rocks, creating metamorphic rocks. The pressures created when two plates collide can metamorphose the sedimentary, igneous, and metamorphic rocks at the plate boundary. As these examples demonstrate, many of the rock cycle processes occur as a result of tectonic activity.

Master Materials List

Originally designed as a program in leadership development, Project Earth Science provides the resources needed to prepare middle-school teachers to lead workshops in topics on Earth science. The leadership workshops have been designed to help teachers convey the content of Earth science through the use of hands-on activities.

We suggest organizing *Project Earth Science: Geology* Activities around four key concept areas, explained further in the Introduction. The four key concept areas are: (1) geological investigation through scientific modelling; (2) plate tectonics, or independent plate motion, and its causes; (3) the relationship between geological phenomena and plate tectonic activity, and; (4) rocks and minerals as products of complex though unified geological processes. To assist workshop leaders, this Master Materials List includes all the equipment necessary for using the Activities that fall within each key concept area.

An Organizational Matrix appears on page 16 and 17. The matrix provides a guide to the way the process-and-inquiry design of *Project Earth Science: Geology* works to support the National Science Education Standards.

Activities:
- GeoPatterns
- All Cracked Up
- Edible Tectonics
- A Voyage Through Time
- Convection
- A Drop in the Bucket
- Seafloor Spreading
- Volcanoes and Plates
- Volcanoes and Hot Spots
- Volcanoes and Magma
- Shake It Up
- Study Your Sandwich, & Eat It Too!

Concept Area I: Geological investigation through scientific modelling

safety glasses or goggles

lab apron

water—hot and cold

scissors

glue or clear tape

masking tape

map of the ocean floor

current world map showing terrain

small Milky Way™ candy bar

hard-boiled eggs, brown or dyed

small, sharp kitchen knife

colored pencils or crayons

narrow- and broad-tipped permanent markers

food coloring (red, blue, and green)

pipette or medicine dropper

paper or styrofoam cups (one with lid)

water basin

large trash can or bucket

tray

clear plastic panwire mesh screen

string with small weight attached

graph paper and pencil

meter stick

non-floating objects, such as rocks, bricks, a bowl

dropping bottle with small neck opening

styrofoam "tectonic plate"

25 cm piece of string

50-100 ml beaker

plaster of paris

hot plate

pan in which to boil water

tongs

two books, same size

shoe box lid

sugar cubes

white, whole wheat, and dark rye bread

jelly

chunky peanut butter mixed with raisins

paper plates

plastic knife

spoon, measuring spoon

clear plastic straws

cleaning supplies, towels

Concept Area II: Plate tectonics

safety glasses or goggles

lab apron

scissors

glue or clear tape

map of the ocean floor

current world map showing terrain

colored pencils or crayons

narrow- and broad-tipped permanent markers

hard-boiled eggs, brown or dyed

small, sharp kitchen knife

small Milky Way™ candy bar

Silly Putty™

hammer

water—hot, cold, and room temperature

tray

board

Activities:
GeoPatterns
All Cracked Up
Edible Tectonics
A Voyage Through Time
Solid Or Liquid?
Convection

food coloring (red, blue, and green)

basin

clear plastic pan

pipette or medicine dropper

paper or styrofoam cups (one with lid)

cleaning supplies, towels

Concept Area III: Geological phenomena and plate tectonics

Activities:
A Drop in the Bucket
Seafloor Spreading
Volcanoes and Plates
Volcanoes and Hot Spots
Volcanoes and Magma
Shake It Up
Rock Around the Clock
Study Your Sandwich, & Eat It Too!
Rocks Tell A Story

safety glasses or goggles

lab apron

large trash can or bucket

wire mesh screen

water-soluble finger paint, food coloring, or ink

water—hot and cold

string with small weight attached

masking tape

permanent marker

crayons or colored pencils

graph paper and pencil

meter stick

non-floating objects, such as rocks, bricks, a bowl

scissors

tape

clear plastic container, such as a shoebox

dropping bottle with small neck opening

styrofoam "tectonic plate"

25 cm piece of string

50-100 ml beaker

300 ml paper or styrofoam cup

plaster of paris

hot plate

pan in which to boil water

tongs

two books, same size

shoe box lid or tray

sugar cubes

pocket pencil sharpener

scrap paper

two pieces lumber, 2.5x12.5x20 cm

aluminum foil

envelopes

newspaper

vise

white, whole wheat, and dark rye bread

jelly

chunky peanut butter mixed with raisins

paper plates

plastic knife

spoon, measuring spoon

clear plastic straws

rock-sample set containing gabbro and basalt, sandstone and conglomerate, slate and shale, marble and limestone, gneiss and granite

cleaning supplies, towels

Concept Area IV: Rocks and minerals

Activities:
Rock Around the Clock
Study Your Sandwich, & Eat It Too!
Rocks Tell A Story

safety glasses or goggles

lab apron

pocket pencil sharpener

scrap paper

crayons

two pieces lumber, 2.5x12.5x20

aluminum foil

envelopes

newspaper

vise

white, whole wheat, and dark rye bread

jelly

chunky peanut butter mixed with raisins

paper plates

plastic knife

measuring spoon

clear plastic straws

rock-sample set containing gabbro and basalt, sandstone and conglomerate, slate and shale, marble and limestone, gneiss and granite

cleaning supplies, towels

Annotated Bibliography

This Annotated Bibliography was compiled by the staff, consultants, and participants of Project Earth Science, and by the NSTA Special Publications editors. It is not meant to be a complete representation of resources in geology, but will assist teachers in further exploration of this subject. The entries are subdivided into the folowing categories:

Activities and Curriculum Projects. This category includes complete curricula, multidisciplinary units, and collections of hands-on activities. Entries identified as "curriculum" are complete units designed to be used as a series of lessons, while entries noted as "activities" are collections of individual activities.

Books and Booklets. Textbooks, story books, and booklets are included in this category.

Audiovisual Materials. Media materials listed in this section include videotapes, filmstrips, and slides.

Instructional Aids. Included in this category are charts, games, photographs, and posters.

Information and References. This category lists additional resources such as: bulletins, bibliographies, catalogs, journals, reference booklets, periodicals, and reports.

State Resources. Each of the 50 United States has its own geological survey, and this category identifies some of the resource materials they are able to provide.

Internet Resources. Starting points for exploration of on-line resources in geology.

Each entry begins with a quick-reference formula in parentheses, a reference, a brief annotation, and an address. The formula consists of:

Category Type. The specific type of item for the category.

Publication Date. The date of publication.

Grade Level. Elementary (K-5), Middle (6-8), High (9-12), and All.

Novostar Designs, Inc.
317 South Main Street
P.O. Box 1328
Burlington, NC 27216-1328
(910) 229-5656
(910) 229-1330 fax
Novostar@aol.com E-mail

Novostar Designs, Inc., offers materials and kits specially designed to accompany activities presented in *Project Earth Science: Geology*.

Activities and Curriculum Projects

Activities in Planetary Geology for Physical and Earth Sciences
(Activities/1990/Middle-High)
NASA publication (EP-179) contains activities that deal primarily with planetary geology. Activities may also be used to illustrate terrestrial geology. (175 pages)

Superintendent of Documents
US Government Printing Office
Washington, DC 20402
(202) 783-3238

Adventures in Geology
(Activities/1989/Elementary-Middle)
Hassard, J. Interdisciplinary lessons in geology with activities and questions.

American Geological Institute
Publications Center
P.O. Box 205
Annapolis, MD 20701
(301) 953-1744
(301) 953-2838 fax

Earthquakes—NSTA/FEMA
(Curriculum/1988/Elementary-Middle)
Presents an interdisciplinary curriculum on earthquakes primarily in the form of inexpensive hands on activities and games. A curriculum guide with black-line masters is included for levels K-5. (169 pages)

National Science Teachers Association
1840 Wilson Blvd.
Arlington, VA 22201
(800) 722-6782
(703) 243-7100 fax

Earth Science Investigations
(Activities/Middle–High/ISBN# 0-922151207-1)
Oosterman, M. and Schmidt, M., eds. Classroom-tested research activities with concepts, vocabulary, and worksheets. (238 pages)

American Geological Institute
Publications Center
P.O. Box 205
Annapolis, MD 20701
(301) 953-1744
(301) 953-2838 fax

Geology: The Active Earth
(Activities/1988/Elementary-Middle)
Part of the Ranger Rick's Naturescope Series. Includes activities and background information on the structure of the Earth, plate tectonics, earthquakes, volcanoes, rocks & minerals, erosion, fossils, and fossil fuels. (68 pages)

Naturescope
National Wildlife Federation
8925 Leesburg Pike
Vienna, VA 22184
(703) 790-4260
(703) 442-7332 fax
http://www.nwf.org/nwf/

Groundwater: Illinois' Buried Treasure Education Activity Guide (Activities/1989/All)
Collection of activities and games dealing with ground water. Activities are designed to help students understand the importance of ground water management. (56 pages)

Department of Natural Resources
524 South 2nd Street
MS211
Springfield, IL 62701-1787
(217) 785-8577
(217) 782-6051 fax

MacMillan
P.O. Box 508
Columbus, OH 43216-0508
(800) 848-9500
(614) 860-1877 fax
http://www.mcp.com/76083392701424/mgr/

Laboratory Manual in Physical Geology
(Activities/1990/High)
Busch, R.M. This manual is set up in lab report format. However, it also contains numerous maps and colored photographs as well as tear-out paper structural models. Laboratory topics include rocks, minerals, topography, water, winds, landforms, earthquakes, and plate tectonics. (216 pages)

Creative Dimensions
P.O. Box 1393
Bellingham, WA 98227
(360) 733-5024
(360) 733-4321 fax

The Mystery of the Far-Flung-Fossils: Investigating Plate Tectonics
(Curriculum/1989/Middle)
Series of activities investigating the theory of plate tectonics. This activity begins with a fossil collection expedition and ties fossil deposits to the theory of continental drift. (17 pages)

GEMS
Lawrence Hall of Science
University of California
Berkeley, CA 94720
(510) 642-7771
(510) 643-0309 fax

River Cutters
(Curriculum/1989/Elementary-Middle/ISBN# 0-912511-67-2)
Part of the "Great Explorations in Math and Science" (GEMS) series. The activities model the passage of geologic time, simulating the effects of thousands of years of erosion. (66 pages)

TOPS Learning Systems
10970 S. Mulino Road
Canby, OR 97013
(503) 266-8550
(503) 266-5200 fax

Rocks and Minerals
(Activities/1989/Middle-High/ISBN# 0-941008-23-1)
Part of the "Task Oriented Physical Science Program" (TOPS). TOPS hands-on activities allow for creativity in problem solving. The manual includes suggestions for organizing and presenting activities, and review and test questions with answers and reproducible student task cards. (36 pages)

Good Apple, Inc.
Box 299
Carthage, IL 62321-0299
(800) 435-7234
(614) 771-7362 fax

Young Scientists Explore Rocks and Minerals
(Activities/1986/Middle/ISBN# 0-866-533-419)
Features activity sheets including experiments, puzzles, cutouts, hidden pictures, and charts. Much emphasis is placed on creativity. Activities may be used in the classroom or for independent study. (30 pages)

Books and Booklets

Earth
(Book-Soft/1985/Elementary/ISBN# 0-816-702-519)
Brandt, K. A short story describing the Earth's physical characteristics, its inner composition, and its crustal movements. It is
written at an elementary reading level.

Troll Associates
Customer Service
100 Corporate Drive
Mahwah, NJ 07430
(800) 526-5289 x1206
(800) 979-8765 fax

Earth in Space
(Booklet/1994/All)
Describes a research project that seeks to improve understanding
of volcanic behavior. Reports on recent activity of 15 volcanos
from around the world, including two in the United States.

American Geophysical Union
2000 Florida Avenue, NW
Washington, DC 20009
(202) 462-6900

Fossils
(Book-Soft/1962/All/ISBN# 0-307-244-113)
Rhodes, F.T. An introductory guide to fossil collection and
geological time. Written as a systematic survey of fossil forms.
(160 pages)

Western Publishing
850 30th Avenue
New York, NY 1 0022
(800) 236-7123
(414) 631-7690 fax

Fundamentals of Geography
(Book-Soft/1990/High/ISBN# 0-697-135-90X)
Doerr, A. and Coling, H. An introductory geography text, cover-
ing land, water, air, life, and their mutual relationships.

Times-Mirror Higher Education Group
2460 Kerper Boulevard
Dubuque, IA 52001
(319) 588-1451
(800) 346-2377 fax

Fundamentals of Geology
Book-Hard/1990/High/ISBN# 0-697-098-060)
Montgomery, C.W. An introductory geology textbook covering a
broad range of topics.

Times-Mirror Higher Education Group
2460 Kerper Boulevard
Dubuque, IA 52001
(319) 588-1451
(800) 346-2377 fax

Geology
(Book-Soft/1972/All/ISBN# 0-307-243-494)
Rhodes, F.T. An introductory guide to geology which discusses
geologic processes, crustal movements, and rocks and minerals.

Western Publishing
850 30th Avenue
New York, NY 10022
(800) 236-7123
(414) 631-7690 fax

Hubbard Scientific
P.O. Box 2121
Fort Collins, CO 80522
(800) 289-9299
(970) 484-1198 fax

Geology Fact Book
(Booklet/1986/Middle-High/ISBN# 0-833-105-728)
Boyer, R.E. Simple introductory manual to geology. This book summarizes basic geological principles.

American Geophysical Union
2000 Florida Ave. NW
Washington, DC 20009
(202) 462-6900

Our Home Planet: A Guide for Secondary School Students
(Booklet/1990/Middle-High)
This booklet is an introduction to geophysics. It outlines the different branches of geophysics and discusses career opportunities.

Western Publishing
850 30th Avenue
New York, NY 10022
(800) 236-7123
(414) 631-7690 fax

Rocks and Minerals
(Book-Soft/1957/All/ISBN# 0-307-244-997)
Zim, H.S. and Shaffer, P.R. Includes general information on rocks and minerals. Techniques on identification, collecting, and classification. Labelling and equipment are also discussed. Note: identification limited to the most common rocks and minerals. (160 pages)

HarperCollins Publishers
10 East 63rd Street
New York, NY 10022
(800) 242-7737
(212) 822-4090 fax

Volcanoes
(Book-Soft/1985/Elementary/ISBN# 0-064-450-597)
Branley, F.M. A "Let's-Read-and-Find-Out" book. This story explains how volcanoes are formed and how they effect the Earth. Written at an elementary reading level.

Audiovisual Materials

American Geological Institute
Publications Center
P.O. Box 205
Annapolis, MD 20701
(301) 953-1744
(301) 953-2838 fax

The American Geological Institute's Videodisc
(Videodisc/All/Item #503)
Includes more than 150 animations of key geological processes, earth-science illustrations, almost 2,000 photographs of geological features, and information on rocks, minerals, volcanic activities, plate tectonics, mountain building, and earthquakes.

Planet Earth Series
Annenburg/Corporation for Public Broadcasting Project
(Video/1986/Middle-High)
The series includes:
The Living Machine (60 min)
Explores the theory of plate tectonics.
The Blue Planet (60 min)
Investigates the ocean depths.
The Climate Puzzle (60 min)
Looks at changes in climate.
Tales from Other Worlds (60 min)
Explores other planets and moons.
Gifts from the Earth (60 min)
Examines Earth's mineral and energy resources.
The Solar Sea (60 min)
Investigates the relationship between the Earth and Sun.
Fate of the Earth (60 min)
Discusses theories about the global consequences of a "nuclear winter" and an "ultraviolet spring."

Corporation for Public Broadcasting
The Annenburg/CPB MultiMedia Collection
P. O. Box 2345
South Burlington, VT 05407-2345
(800) 532-7637
(802) 864-9846 fax

Born of Fire
(Video/1983/Middle-High)
This video illustrates the results of the movement of Earth's huge crustal plates. It features worldwide footage of earthquakes and volcanoes.

National Geographic Society
Educational Services
17th and M Streets NW
Washington, DC 20036
(800) 368-2728
(301) 921-1575 fax

The Earth Has a History
(Video/1989/Middle-High)
A Geological Society of America (GSA) study module demonstrating the concept of geologic time. Supplemental instructor's materials include lesson plans and black-line masters. (20 min)

Geological Society of America
3300 Penrose Place
P.O . Box 9140
Boulder, CO 80301
(800) 472-1988
(303) 447-1133 fax

Our Dynamic Earth
(Video/1979/Middle-High)
This video presents the theory of plate tectonics and details the different types of plate movement. (23 min)

National Geographic Society
Educational Services
17th and M Streets NW
Washington, DC 20036
(800) 368-2728
(301) 921-1575 fax

Educational Images Ltd.
P.O. Box 3456
Elmira, NY 14905
(607) 732-1090
(607) 732-1183 fax

Plate Tectonics and the Spreading Sea Floor

(Slides/1989/Middle-High/Item #310)

This 20-slide set illustrates the geologic principals of plate tectonics and sea floor spreading. A companion guide gives background information and details on each slide. (20 slides)

Instructional Aids

Focus Media, Inc.
839 Stewart Avenue
P.O. Box 865
Garden City, NY 11530
(516) 931-2500
(516) 931-2676 fax

The Earthquake Simulator

(Software/1986/Middle)

This software package includes an instructor's manual, lesson plans, student workbook and disks (Apple II). Lessons include: Plate Movement, Earthquake Waves, and Faults and Folding. (27 pages)

Apple format only:

3.5" disk 0-780-104-96X   3.5" site license 0-780-104-978
3.5" lab pack 0-780-104-986   5.25" site license 1-555-147-992
5.25" lab pack 1-555-144-039 5.25" disk 1-555-141-943

U.S. Geological Survey
345 Middlefield Road
Menlo Park, CA 94025
(415) 853-8300
(415) 329-5130 fax

Fault Motion

(Model & Software/1989/High)

This open-file report (89-640A) describes how to construct a paper model showing the motion of the San Andreas fault. Accompanying software is available for Macintosh and Silicon Beach SuperPaint graphics software.

American Association of Petroleum Geologists
P.O. Box 979
Tulsa, OK 74101
(918) 584-2555
(918) 560-2652 fax

Geological Highway Maps

(Maps/All)

The American Association of Petroleum Geologists has produced geologic maps of 12 regions of the U.S. Maps highlight the geologic history of each region and contain a brief discussion of fossils, minerals, and gemstones.

Tom Snyder Productions, Inc.
90 Sherman Street
Cambridge, MA 02140
(617) 926-6000
(617) 926-6222 fax

Geoworld: A Living Database

(Software/1986/Middle-High)

This software package includes disks (Apple II) and an instructor's manual. Fifteen accurate natural resource databases are used to simulate geological exploration. The database may also be used for a game.

The Topo Kit
(Mapping kit/1989/All)
A mapping tool used to create topographic profiles from maps
provided from direct site measurements.

Novostar Designs, Inc.
P.O. Box 1328
Burlington, NC 27216-1328
(910) 229-5656
(910) 229-1330 fax

Information and References

Earthquake Educational Materials for Grades K-12
(Bibliography/1991/All)
Resource for educators interested in earthquake education.
Curricula, books, articles and software are included in this
annotated bibliography. Citations include reading levels and
length of books whenever possible.

National Center for Earthquake Engineering
Research
State University of New York at Buffalo
Red Jacket Quadrangle
Buffalo, NY 14261
(716) 645-3391
(716) 645-3399 fax

Earthquake Information Bulletin
(Periodical/Bi-monthly/All)
The bulletin provides current information on earthquakes and
seismic activity.

Superintendent of Documents
U.S. Government Printing Office
Washington, DC 20402
(202) 512-1800
(202) 512-2250 fax

Earth Science Education Resource Directory
(Information and references/Annual/All)
Describes educational resources from over 170 earth-science
organizations. Includes audio-visual, computer-based, displays,
manipulatives, workshops/speakers. (152 pages)

American Geological Institute
4220 King Street
Educational Images Ltd.
P. O. Box 3456
Elmira, NY 14905
(607) 732-1090
(607) 732-1183 fax

Earthquakes and Volcanoes
(Periodical/Bimonthly/All)
This publication provides current information on earthquakes,
seismology, volcanoes, and related natural hazards.

Superintendent of Documents
U.S. Government Printing Office
Washington, DC 20402
(202) 512-1800
(202) 512-2250 fax

Eruptions of Hawaiian Volcanoes: Past, Present, and Future
(Booklet/1987/All)
Tilling, R.l., Heliker, C., and Wright, T.L. This pamphlet presents selected highlights from Hawaiian volcanoes past and
present.

U.S. Geological Survey
Federal Center
Box 25046
Denver, CO 80225
(303) 202-4200
(303) 202-4188 fax

U.S. Geological Survey
Federal Center
Box 25046
Denver, CO 80225
(303) 202-4200

General Interest Publication of the U.S. Geological Survey
(Catalog/All)
A catalog listing various publications of geological interest.

American Geological Institute
4220 King Street
Alexandria, VA 22302-1502
(703) 379-2480
geotimes@jei.umd.edu e-mail

Georef
(Database/All)
More than 1.9 million references to geoscience journal articles, books, maps, conference papers, reports, theses. Available through CD-ROM, Internet, on-line, network license, or in print.

American Geological Institute
Publications Center
P.O. Box 205
Annapolis, MD 20701
(301) 953-1744
(301) 953-2838 fax
http://agi.umd.edu/agi/agi.html

Geotimes
(Periodical/Monthly/All/ISSN# 0-016-855-6)
Periodical published monthly by the American Geological Institute. This publication contains geological current events (e.g., earthquakes, environmental problems, volcanic eruptions, important developments in Earth sciences), book and film reviews, and information about books and teaching aids.

The National Association of Geology Teachers
Bob Christman
P.O. Box 5443
Bellingham, WA 98227-5443
(360) 650-3587
(360) 650-7295 fax

Journal of Geological Education
(Periodical/Yearly/High/ISSN#0-022-136-8)
This periodical is published by the National Association of Geology Teachers. Written mainly for college teachers, but some of the teaching materials and suggestions can be used by middle and high school teachers. (Some articles are specifically for pre-college teachers.) Includes geological book, film and museum reviews.

Peri Lithon Books
5372 Van Nuys Ct.
PO Box 9996
San Diego, CA 92109
(619) 488-6904

Peri Lithon Books
(List/Bimonthly/All)
An alphabetical listing of books on gems, precious stones, gemology, minerals, geology, etc.

TOPS Learning Systems
10970 S. Mulino Road
Canby, OR 97013
(503) 266-8550
(503) 266-5200 fax

Pressure
(Curriculum/1979/Middle-High/ISBN# 0-941-008-86X)
Part of the Task Oriented Physical Science (TOPS) program. This module includes 20 activities that illustrate the fundamental concepts of pressure.

ScienceNews
(Periodical/Weekly/Middle-High)
A science digest periodical that often contains articles pertaining to geology and recent geological findings.

Editorial and Business Offices
Science News
1719 N Street, NW
Washington, DC 20036
(202) 785-2255

Selected Books on Geology and Related Subjects
(Bibliography/Annual/All)
List of non-technical books in the field of geology.

Geological Inquiries Group/U.S.G.S.
907 National Center
Reston, VA 22092
(703) 648-4000
(703) 648-5548 fax

Solid-Earth Sciences and Society
(References/All/ISBN# 0-309004739-0)
National Research Council report describing the state of the earth sciences, including research issues and societal needs. (346 pages)

American Geological Institute
Publications Center
P.O. Box 205
Annapolis, MD 20701
(301) 953-1744
(301) 953-2838 fax

U.S. Geological Survey Catalog of Maps
(Catalog/Annual/All)
U.S. Geological Survey catalog of maps including: topographic maps, photo image maps, satellite image maps, geologic maps, hydrologic maps, maps of planets and moons, land use maps, and more.

U.S. Geological Survey
Federal Center
Box 25406
Denver, CO 80225
(303) 202-4200
(303) 202-4188 fax

U.S. Geological Survey Teacher's Packets
(Packet/Annual/All)
Available on request. Submit your request on school letter head, indicating subjects and grade level taught. Available packets include:
• Teachers Packet of Geological Materials (Packet of teaching aids compiled for high school and college instructors).
• Selected Packet of Geological Teaching Aids (Packet of teaching aids compiled for elementary and middle school teachers).

U.S. Geological Survey
Federal Center
Box 25406
Denver, CO 80225
(303) 202-4200
(303) 202-4188 fax

State Resources

Each of the 50 United States operates its own geological survey, and all states publish a wide variety of materials relating to geology, geological phenomena, and the relationship between those phenomena and state citizens, such as emergency preparedness guides. Below is a sample of the kinds of resources your state may provide.

These publications are put out by the North Carolina Geological Survey
North Carolina Department of Environment, Health, and Natural Resources
P.O. Box 27687
Raleigh, NC 27611-7687
(919) 733-2423
(919) 733-0900 fax
http://www.ehnr.state.nc.us/EHNR/

A Geologic Guide to North Carolina's State Parks
(Bulletin /1989/All)
A geological guide (Bulletin #91) to 33 state parks and recreational areas in North Carolina. (69 pages)

Geologic Map of North Carolina
(Map/1985/All)
A color map, 44 x 66" with a 1,500,000 scale.

Fossil Collecting in North Carolina
(Bulletin/1988/All)
Carter, J.G., et al. North Carolina Department of Natural Resources and Community Development, Geological Survey Section, Bulletin 89. This bulletin lists and describes 34 fossil collecting sites in North Carolina. (89 pages)

Mineral Collecting Sites in North Carolina
(Circular/1978/All)
Wilson, W.F. and McKenzie, B.J. North Carolina Department of Environment, Health, and Natural Resources, Geological Survey Section, Information Circular #24. A history of gem mining in North Carolina. A list of gem collecting sites and maps are provided.

Carolina Academic Press
700 Kent Street
Durhem, NC 27701
(919) 489-7486
(919) 493-5668 fax

North Carolina: The Years Before Man
(Book-Hard/1991/Middle/ISBN# 0-890-894-000)
Beyer, F. This book recounts the geologic history of North Carolina. It begins with a brief history of the planet and continues through present day.

Internet Resources

The Internet is filled with resources pertaining to geology that are appropriate for student use. Key-word searches yield the best results, even simple words such as "volcano" and "earthquake." The images on the cover of this book were obtained through the Internet, and it is currently the best and quickest resource for the most up-to-date satellite imaging of global geological phenomena. Examples of sites include:

American Geological Institute
http://jei.umd.edu/agi/agi.html

Covis Geosciences
http://www.covis.nwu.edu/Geosciences/index.html

Earthed: Earth Science Education Resources
http://www-hpcc.astro.washington.edu/scied/earth.html

The Electronic Volcano
http://www.dartmouth.edu/pages/rox/volcanoes/elecvolc.html

Geological Servers on the Web
http://www.geology.utoronto.ca/geological/geological.html

Geological Society of America
http://www.aescon.com/geosociety/index.htm

Geology in the Classroom
http://agcwww.bio.ns.ca/schools/school-index.html
Includes geologists available to answer student and teacher questions, an earth science resource catalog, Canadian Geoscience Education Network, and information about workshops for teachers.

General Earth Science Resources
http://unite2.tisl.ukans.edu/Browser/UNITEResource/RNatural_Science483.html
K–12 math and science resources: instructional software, lab activities, lesson plans, etc.

Geoscience K-12 Resources
http://www.cuug.ab.ca:8001/~johnstos/geosci.html

Hotlist—Geology
http://sln.fi.edu/tfi/hotlists/geology.html
Resources in general geology, volcanoes, earthquakes, rocks and minerals, teacher resources including lesson plans.

Illinois State Geological Survey
http://denr1.igis.uiuc.edu:/isgsroot/dinos/dinos_home.html

Indiana Geological Society
http://www.indiana.edu/~igs/aboutigs/aboutigs.html

Internet Earth Science Resources
http://gbyerly.geol.lsu.edu/geology/sources.html

Incorporated Research Institutions for Seismology (IRIS) Data Management System Home Page
http://dmc.iris.washington.edu/

Joint Education Initiative
http://jei.umd.edu/jei.html
Includes K-12 information and information about teacher workshops.

Michigan Technological University Volcanoes Page
http://www.geo.mtu.edu/volcanoes/

NASA EOS IDS Volcanology Team
http://www.geo.mtu.edu/eos/

National Oceanic and Atmospheric Administration's (NOAA) National Geophysical Data Center
http://www.ngdc.noaa.gov/ngdc.html

National Science Foundation
http://www.nsf.gov/

New Hampshire Geological Society
http://kilburn.keene.edu/EarthScience/NHGS/NHGS.html

National Science Teachers Association
http://www.nsta.org

On-Line Resources for Earth Scientists—Geological Resources
http://www.calweb.com/~tcsmith/ores/geology/index.html

Smithsonian Gem and Mineral Collection
http://galaxy.einet.net/images/gems/gems-icons.html

Southern California Earthquake Center
http://scec.gps.ca;tech.edu/

Thomas Jefferson High School for Science and Technology
http://www.tjhsst.edu/
Includes information about the school's geoscience technology lab and geoscience department.

University of Arizona Department of Geosciences: K-12 Geoscience Education
http://www.geo.arizona.edu/K-12/

University of Nevada—Reno Seismological Laboratory
http://www.seismo.unr.edu/

U.S. Department of Education
http://www.ed.gov/

U.S. Geological Survey
http://info.er.usgs.gov/

Volcanoes and Earthquakes and Geoscience
http://sharkbait.arl.psu.edu/Scott/058.html

Volcano World
http://volcano.und.nodak.edu